Aegyptus et Pannonia 1.

Acta Symposii anno 2000

Éditeurs: Hedvig Győry – Zsolt Mráv
Publication du Comité de l'Égypte Ancienne de l'Association Amicale Hongroise-Égyptienne (MEBT – ÓEB)

Cette publication a été réalisée par l'aide de Ibisz Bt.

Couverture design: Balázs Horváth
Sur la couverture:
le vase de Egyed
(Musée National de Hongrie, Budapest)
Photo de couverture: András Dabasi
Réalisation technique: Aquila Design
Imprimerie: Managerpress
ISBN: 963 204 851 2

Aegyptus et Pannonia 1.

Acta Symposii anno 2000

Budapest – 2002

TABLE DES MATIÈRES

H. Győry – Zs. Mráv: Préface 7-11
Adél Bugán: The face of Marcus Aurelius 13-20
Kornelia Giunio: Egyptian Gods on the Eastern Adriatic Coast 21-63
Hedvig Győry: Les terres cuites d'Harpocrate-au-faucon-sur-le-pilier 65-97
Peter Hubai: Religionsgeschichtliches zur pannonischen Isis 99-122
László Kákosy: Isis in Rom 123-127
Tamás Mezős: A new concept for the Reconstruction of the Iseum of Szombathely 129-145
Zsolt Mráv: A Ring Decorated with the Bust of Sarapis in the Collection of the Hungarian National Museum 147-162
Klára Póczy: Veränderungen des Hintergrundes der ägyptischen Kulte in Pannonien im 1.-4. Jh. 163-172
Ottó Sosztarits: Topographische und chronologische Problemen des Iseums von Savaria 173-174.
Ádám Szabó: Eléments d'interprétation des autels de Sphinx de Pannonie 175-191
Tihamér Szentléleky: Einige wichtige Fragen im Zusammenhang mit der Bearbeitung des Iseums von Savaria 193-199
István Tóth: Apuleius "pannonicus" 201
Edith Varga: Monuments égyptiennes en Hongrie 203-213
Abrèviations et bibliographie 215-248
Planches 1-32

PRÉFACE

En entendant le mot Égypte, de nombreuses associations viennent à l'esprit de l'homme d'aujourd'hui. On se souvient de l'empire florissant des pharaons, de leurs activités monumentales et des monuments merveilleux, de la ville d'Alexandrie avec sa bibliothèque fabuleuse, de Cléopâtre, la célèbre reine, de l'église chrétienne des coptes avec le monde légendaire des martyrs et des ermites, des mosquées et des petits objets d'art, de la religion islamique, des tendances politiques de l'Égypte moderne, des paysages particuliers de la vallée du Nil, de la faune et de la flore somptueuses de la mer Rouge, du monde éteint mais toutefois vivant du désert, du Wadi Soura, découvert pour l'ère moderne par l'expédition de László Almássy, de l'Aegyptopithecus, qui représente une étape importante sur la route du développement des hommes, la momification, les pyramides, le phare de Pharos, la foule fourmillante de Khan El-khalili... Sa variété et sa diversité ont fait de l'Égypte le centre d'intérêt du monde, non sans raison.

Mais pas seulement aujourd'hui. Les Hongrois avaient des contacts directs ou indirects avec les Egyptiens déjà au temps du premier roi, Saint Etienne. Une des histoires légendaires de la vie de l'évêque Saint Gellért, où il exalte le chant de la servante travaillant avec la meule, ressuscite un motif trouvé dans l'enseignement du Ptahhotep de l'Ancien Empire. Parmi les joyaux du couronnement hongrois le cristal des montagnes colossal du sceptre est aussi d'origine égyptienne. Au cours du Moyen Age, les rois hongrois ont maintes fois envoyé des légats à la cour égyptienne et plusieurs pèlerins se rendant en Terre Sainte y sont aussi allés pour trouver le mont Moïse ou l'abri de la famille sainte à Héliopolis. Les noms de György Raszinyai Huszti et Gábor Pécsváradi sont bien connus dans ce contexte. Puis beaucoup d'enfants enlevés et élevés pour être des janissaires pendant l'époque de la domination ottomane ont été installés en Égypte. La population nommée "magarab" s'est développée à partir des soldats de ces garnisons, à travers des mariages avec des femmes arabes locales. Ils sont venus en Égypte contre leur volonté, mais d'autres s'y sont rendus de leur plein gré, comme Lukács Gyertyánffy qui a visité plusieurs temples en Haute-Égypte et en Nubie, ou Sándor Kõrösi Csoma qui voulait étudier les chroniques arabes pour obtenir des informations sur nos ancêtres. Après la révolution et la lutte pour l'indépendance hongroise en 1948/49, plusieurs des émigrants hongrois ont trouvé asile en Égypte, et plus tard la colonie hongroise a été élargie par des commerçants, des artisans ou bien des artistes nouveau-venus. Au cours des dernières années, Péter Gaboda (Musée des Beaux-Arts) a effectué des recherches poussées dans ce domaine. Tant que les liaisons diplomatiques et commerciales se développaient, outre les voyageurs diplomatiques et commerciaux, les groupes de visiteurs étaient de plus en plus nombreux. Ils venaient pour voir les monuments fabuleux de leurs propres yeux, pour les étudier ou encore pour des raisons de santé. En effet, pour certaines maladies, en particulier pour celles liées à des infections pulmonaires, les médecins européens des siècles précédents conseillaient de longs voyages en Méditerranée.

Le résultat se manifestait aussi par l'achat d'antiquités égyptiennes par ceux qui étaient installées en Egypte, par les visiteurs et également dans le cadre de la mission archéologique inaugurée par Philipp Back. Comme ils racontaient leurs aventures avec beaucoup de plaisirs et que nombreuses de leurs acquisitions sont parvenues en Hongrie, le public hongrois s'est aussi mis en contact direct avec les différentes cultures anciennes de l'Égypte. D'abord par l'exposition centrale du Musée National Hongrois qui en a présenté quelques pièces – et les collections privées, comme p. ex. celle de Fehérváry-Pulszky, puis par une exposition dans une salle indépendante du bâtiment de la Collection Ethnographique du Musée National (1912) ont permis de présenter des matériaux trouvés lors des fouilles, à Charouna et à Gamhoud. En même temps beaucoup d'objets de l'Égypte ancienne étaient dispersés dans d'autres collections, mais ils prenaient en général la poussière dans des dépôts. C'est la raison pour laquelle Ede Mahler a insisté pour rassembler ces différentes pièces dans une grande collection égyptienne. Et vraiment, dans les plans de construction du Musée des Beaux-Arts, une exposition égyptienne était prévue, mais le projet n'a pu être mené à bien qu'après la loi sur les musées de 1934. Pourtant cela ne signifie pas que toutes les antiquités égyptiennes y sont exposées – des musées régionaux présentent également quelques pièces égyptiennes.

La culture et l'art spécifique de l'Égypte Ancienne, hellénisés dans les apparences, étaient bien connus sur le territoire de la Hongrie, surtout en Pannonie, même à l'époque impériale romaine. La connaissance ne parvenait pas seulement aux habitants de la province par le biais des soldats qui avaient servi en Égypte ou des commerçants qui étaient en contact avec l'export des marchandises égyptiennes – surtout du blé et des "touristes" qui allaient voir ce pays fabuleux déjà en ce temps-là. La tendance égyptienne à la mode au 2e s. apr. J.-C n'a pas non plus été le seul moyen de faire connaître cette culture dans la province. Les religions égyptiennes popularisées, répandues dans toute la Méditerranée, en premier lieu par le culte des dieux alexandrins ont eu un rôle important dans ce processus. Au 1e s. apr. J.-C., la déesse Isis était très appréciée en Pannonie. Sa popularité prenait racine dans ses qualités bienfaitrices, et par sa position au-dessus du destin; elle pouvait aider dans toutes sortes de situations fâcheuses. Sa personnalité a été repoussée à l'arrière-plan seulement au cours du 3e s. apr. J.-C. et par Sérapis, son mari mythique, qui jouissait de l'assistance officielle de la maison impériale surtout dans le monde de l'armée de Pannonie. Les monuments de cette religion riche et multicolore n'ont donc pas seulement été découverts sur les territoires de l'Empire Romain, mais aussi en Pannonie, ce qui signifie que la culture Égyptienne et ses modifications sont aussi devenues parties intégrantes de notre héritage culturel.

Des objets précieux peu communs ont attiré l'attention de la recherche hongroise sur de nombreux objets égyptiens ou inspirés de la culture égyptienne découverts en Hongrie. On distingue par exemple un ensemble de pièces de vaisselle utilisées pour le culte, publié par Antal Hekler, avec des incrustations d'or et d'argent qui représentent les dieux égyptiens. Ces pièces ont été trouvées à Egyed. Une impulsion semblable a été donnée à la recherches des monuments de la culture égyptienne par la découverte du sanctuaire de Savaria, décoré par des marbres représentant des dieux égyptiens, résultat des recherches de Tihamér Szentléleky. Mais l'autel découvert près du lac Balaton ou les reliefs d'influence égyptienne trouvés à Gorsium, de son nom romain, et analysés en détails par Vilmos Wessetzky, sont aussi très importants. Vilmos Wessetzky est aussi la personne qui a recuilli et analysé la première fois l'ensemble de ces monuments et les a fait connaître aux archéologues hongrois et étrangers dans le premier volume de EPRO qui allait devenir la série monumentale internationale des antiquités Égyptiens et d'influence égyptienne dans le monde antique. L'œuvre de Aladár Dobrovits, autre grand personnage dans l'histoire hongroise de l'égyptologie, a suivi, il a rassemblé les antiquités trouvées à Budapest, Aquincum romain, et s'est également penché sur les origines des objets égyptiens des tribus "barbares". András Alföldi a contribué en premier lieu à la compréhension des cultes d'Isis, et László Castiglione, à la compréhension de celui de Sérapis. Mais il y a aussi beaucoup de chercheurs contemporains qui sont fort interessés aux monuments et tendences égyptisants de Pannonie.

En raison de ce large intérêt, le Comité de l'Égypte Ancienne de l'Association Amicale Hongroise-Égyptienne et le Musée National Hongrois ont décidé d'organiser un symposium intitulé "Aegyptus et Pannonia" qui offre la possibilité de présenter de nouveaux résultats hongrois dans ce domaine. Ce volume contient des exposés présentés en 2000, transformés en articles et un article portant sur le bord de l'Adriatique, qui est en retard, mais offre un matériau de comparaison intéressant.

Enfin il ne reste que le souhait des éditeurs d'exprimer ici leur reconnaissance envers tout ceux qui ont contribué à l'organisation du symposium et à l' apparition de ce volume, qui devrait bientôt être suivi d'un prochain contenant les exposés du Symposium qui aura lieu en 2002.

Hedvig Győry – Zsolt Mráv
éditeurs

THE FACE OF MARCUS AURELIUS

Marcus Aurelius, a Roman emperor, is known more intimately before the posterity than any of his successors, for he left behind manifestations written in his own hand. Reading his confessions, we are presented with a picture of a man who lived by Roman stoic ethics through the problems of his own life, rather than professing principles as a philosopher. The head from a monument found in Dunaszekcső[1] was discovered almost three decades ago and it can be considered one of the best portraits of Marcus, because it depicts not the emperor, but the man known from the "Meditations" (Pl. 1/1-2).

This depiction seems so intimate that the location of the find is hardly believable. In the area of Dunaszekcső, Lugio in ancient times, there was a small military fort of medium strategic importance. The castellum was built in the place the find was located, on the so-called Várhegy (a small hill which faces the Danube). Here, a shrine to the Emperor cult may have stood. Is it possible to believe that a monument of Marcus Aurelius, portrayed with true realism and free of any poses, was placed on the Pannonian front, just to show the emperor engrossed in his thoughts, refined with stoic philosophy, in order to effect only the local soldiers only with his personal charm? Is it possible to believe that the natives were so well acquainted with the emperor, who indeed often stayed in Pannonia, that the sculptor was able to completely by-pass all authoritarian poses?

Depictions on the obverse of contemporary coins are

1 KOVÁCS 1984, 89-, pl. 44-45; FITTISCHEN – ZANKEN 1985, 73, note 9d; MELUCCOVACCARO 1989, 80, fig. 42; CHAMAY 1996, 129-136.

schematic and yet still enhance the characteristic features of Marcus Aurelius' face. On the basis of this kind of comparison we can declare the shape of the hair on the Dunaszekcsõ head to be trustworthy. His locks curl into finger-breadth spirals in a natural manner and curve aside horizontally from the middle of his forehead. His uncombed beard may have curled similarly to his hair, but generally it seems to be more regular and trimmed on other portraits where the beard is always shown curving on both sides, rounded at the bottom, pointed in the middle (Pl. 1/3). On the Dunaszekcsõ depiction the beard is short above the ears and on the face and long and not curved under the chin. The beard falls in a natural curly fashion, clustering into two sections. The wrinkled forehead and the attentively lifted, thick and long eyebrow features may be seen on the coins, too. It is probable that this was his most typical facial expression, which was transformed into the officially accepted portrayal. Reading "Meditations" one may recognize in several of the ideas that well-intentioned attentiveness was part of his conscious manner. We can read in the first chapter of the work: *Παρὰ Σέξτου τὸ εὐμενές· καὶ τὸ παράδειγμα τοῦ οἴκου τοῦ πατρονομουμένου:... και τὸ σεμνὸν ἀπλάστως καὶ τὸ στοχαστικὸν τῶν φίλων κηδεμονικῶς* ...[2] The Dunaszekcsõ head is the only depiction on which the look is not directed towards the outer world. The eyebrows are raised as on several other coin portraits, but the eyes do not focus on anything in particular, looking vacantly into space. His facial bones are wide and a double wrinkle crosses his forehead. His sharp-featured nose is symmetrical. The face is unwrinkled, the muscles are sensitive and do not show any signs of bitterness or worry, fatherly goodwill, authority or soldierly discipline as on other portrait statues of him.

Marcus' earliest portrait, from the age of 10-13, is known

2 Marcus Aurelius, Meditations, I.9. Translation by I. Edman: "*...from Sextus, good humor and the example of a family governed in a fatherly manner ... and top tolerate ignorant persons and those friends who form opinions without consideration...*"

from the Musei Capitolini in Rome. Later depictions are found on coins. On coins dated to 140 AD, when he was 19, the face is not bearded while the nose is large like many teen-agers' (Pl. 1/4-5).[3] At the age of 30 he had a straight combed beard and moustache and his hair conformed to his forehead (Pl. 1/6).[4] On the basis of comparing these features and the known depictions, it was possible to include the marble head held in the Vatican Museum among the portraits of Marcus.[5] Previously it was considered to be a portrait of an unknown person until the Dunaszekcsõ head was found (Pl. 2/2).

The nudity of the figure made the identification doubtful. We cannot suppose that Marcus had an unclothed monument made of himself. But the similar appearance of the faces forces us to revise the possible circumstances of its origin. The depiction of the body hardly differs from the copies of Polykleitos' Doryphoros (Pl. 2/1).[6] Both are statues of perfect athletes without any deformities, without any personality. The slender muscles are typical of 20-25-year-old men, but one may recognise the features of a 35-year-old or older acknowledged men of mature character. Obviously the sculptor did not model the actual body of the emperor, but rather kept an accurate copy of Polykleitos' masterpiece in view during carving work. The Doryphoros, on which its creator wrote the Kanon, was the Greek etalon for well-trained Roman age sculptors. The body seems to be an expression of the ideal of the Classical Greek sculpture while the head is an individual, portrait type. One might think that Marcus Aurelius knew the sculptor personally as one who perhaps made the statue monument as a present by casting it from bronze, employing the same method used by Polykleitos, because he was familiar with his way of thinking, his sympathy for

3 Sest Roma 140-144 in: BMC 1397; Hungarian National Museum, Budapest: inv. no. MNM 110/1896.12.

4 Sest Roma 148-149 in: BMC 1849; Hungarian National Museum, Budapest: inv. no. MNM k15557.

5 LIPPOLD 1956, III.1. pl. 54,62.590.

6 STEUBEN 1973.

Athenian stoic philosophy and expertise in Greek rhetoric. Only the marble copy of this monument has then survived. Undoubtedly, for educated people in the milieu of Marcus, the message and the art-historic essence was more obvious than the nudity of the figure. This piece mostly differs from other portraits of the emperor in its personality although, in the end, it would not have been proper to exhibit it as a tool of imperial propaganda for the masses. Just as the "Meditations", which was written in Greek, was understandable only for those who could read Greek, that is to say, the best educated social classes. That is why we think that the face of the writer of "Meditations", free of any ruler's poses, is the same as that also found in the Vatican Museum.

The similarity of the two faces could be explained by searching for the style of a single sculptor for both heads. However, in terms of the subject of the research, it is more useful to examine the differences between the depictions. The model for the Dunaszekcsõ portrait seems to be of an at least ten years older philosopher, gazing out over the everydayness of the world. His closest friends sometimes saw Marcus Aurelius like that in his immersed moments, but it is hard to explain why this portrait, formed with such surprising realism, would have been placed in Lugio as part of an Emperor cult monument for the soldiers. It is known that in his speeches, the rebel Avidius Cassius, who wanted to proclaim himself emperor in 175 AD, pointed out that Marcus' philosophic way of thinking made him unfit to govern and command.[7] It seems to be a daring idea to represent the emperor with his most vulnerable attributes, and to place this portrait on the Pannonian front, in the area of a military fort.

Using the lost wax process, goldsmiths of that time were able to cast a number of identical bronze statues. Perhaps, several such casts were made and ultimate reached Lugio. One may think that these heads survived accidentally, but it is impossible to believe that the fact that it came to light in Pannonia is actually

7 Cass. Dio LXXI, 17.

accidental, in the very province where the emperor stayed during the last period of his rule.

Marcus stayed twice in Pannonia for long periods. The first time was from 172-174 AD and the second time from 177 AD until his death. The most frequently discussed event of his Pannonian stay was the “Rain Miracle”.[8] The reliable description of the officially accepted history of the propagandistically aimed legend can be read in Cassius Dion. According to the story, north of the Danube, in the Barbarian territory, a battle of crucial importance was fought. During this skirmish, the Barbarians surrounded the Roman soldiers, who thirsted in the summer heat until... *τοι λόγος ἔχει Ἀρνοῦφίν τινα μάγον Αἰγύπτιον συνόντα τῷ Μάρκῳ ἄλλους τέ τινας δαίμονας καὶ τὸν Ἑρμῆν τὸν ἀέριον ὅτι μάλιστα μαγγανείαις τισὶν ἐπικαλέσαστθαι καὶ δἰ αὐτῶν τὸν ὄμβρον ἐπισπάσασθαι.*[9] So, the Egyptian magician called Hermes aeiros, that is Mercury, and made the rain fall. The soldiers drank the water coming from the sky so the Roman army recovered and defeated the Barbarians. This scene can be seen on the relief of the Aurelian column in Rome (Pl. 2/3).[10]

Because we know his name and office from an Aquileian inscription we may suppose that Arnuphis was a real historical person.[11] István Tóth proved, using several contemporary depictions on coins and inscriptions as verification, that from the 170s AD the Egyptian gods of the family became propagandistic political and religious tools during the Marcomann wars and that the event prepared the way for the spread of the Egyptian

8 BIRLEY 1966, 237-242.

9 Cass. Dio LXXII, 8.4. Tanslation by E. Casy: “... *there is a story to the effect that Arnuphis, an Egyptian magician, who was a compagnion of Marcus, had invoked by means of enchantments various deities and in particular Mercury, the god of the air, and by this means attracted the rain.* “

10 CHAMAY 1996, 84-89, fig. 2.

11 SIRIS 613, L’Année Épigraphique 1934, 245.

12 TÓTH 1976, 45-51.

cults along the Danube.[12]

It is obvious that the mysterious believes involved in the worship of Egyptian gods was foreign to the writer of the "Meditations". Still, he permitted them to be worshipped as helpers during the Pannonian wars. As a background to these events, we would like to mention the religious-political reasons, which explain why the emperor chose the Egyptian gods and the Greek Hermes, who was worshipped as Mercury by the Romans, as supporters. Also, it needs to be explained why this cult spread during the next decades. Seventy years before the "Rain Miracle", in an ethnographical analysis of the Germans, Tacitus wrote that: *Deorum maxime Mercurium colunt, cui certis diebus humanis quoque hostiis litare fas habent....Pars Sueborum et Isidi sacrificat; unde causa et origo peregrino sacro, parum comperi, nisi quod signum ipsum in modum liburnae figuratum docet advectam religionem.*[13]

According to the Roman interpretation, the main god of the Germans seems to have been Mercury while one of their goddesses was Isis. The main gods of the Germans became known as Mercury and Isis. Before the Marcomann war, in the 160s AD, Marcus fought against these Germanic tribes. The "Rain Miracle" made it obvious to the soldiers that the Egyptian Isis, identified with the German Isis, had helped the emperor and his soldiers to defeat the enemy, therefore expressing her support for the Roman Empire in its fight against the Barbarians. It is clear that religious-political causes had nothing to do with the Emperor's view of life, but rather that the Roman view of Egyptian gods had outgrown the original Underworld and the Ptolemaios era concept of gods providing a Hellenistic type of royal power. At the end of the 2nd century AD in Apuleius' "Metamorphosis", Isis has become a sincretistic god and her cult mirrors a pantheistic view of life. In "Meditations", Marcus mentions his thoughts on death several times, which are close to the view, expressed by the story of Egyptian gods in mysteries and myths. The study

13 Tacitus, Germania, cap.IX.

of the Egyptianising phenomena of the Marcus era obviously should be done with special care. But perhaps the similarity between the Roman period Serapis depictions (Pl. 2/4.)[14] and the personal portraits of Marcus Aurelius should not be considered paradoxical.

The depictions of the Egyptian god display several similarities. Their identical features are related to a well-known cult monument, which served as a model for the others. This model was probably placed in the Sarapeion in Alexandria. We know from descriptions from the 2nd century BC, that the cult monument in the shrine was made by Bryaxis and it was carried from Sinope to Alexandria at the request of Ptolemaios Soter. According to László Castiglione, by the time of the Roman Imperial period, it was not Bryaxis' monument but another similar one, which stood there. This new monument expressed a new way of thinking about the god's view, placing more emphasis upon the unearthly nature of Serapis' character. This later work can be best imagined as a mixture of similar features from the copies: Serapis has long, uncombed curly hair. A deep wrinkle crosses his forehead, which is covered by a 5-branched lock of hair. His beard is individually combed and curves into two curls. The face has an aimless look characterised by a mysterious reverie. The tranquillity of the face makes it appear almost dead-pan. The most apparent features are exactly reminiscent of individual characteristics of the Dunaszekcsõ Marcus portrait. Only the 5-branched lock is missing, something which could have been omitted from a portrait. Though it was foreign to Marcus' personality to portray himself as having the character of any gods, heroes or mythological people, we can still imagine that during the events of the 170s AD, he permitted the sculptor to portray him with a slight ambiguity as an ancient king from an Egyptian myth, thus, allowing the monument to refer to the support given by Egyptian gods to the soldiers. This masterpiece, made in an Italian workshop, was sent to Lugio because both its realistic pre-

14 LIPPOLD 1956, no. 549.

sentation and the suggestion it gave of gods' power increased the soldiers' trust in the emperor, at the same time that a belief in wonders grew during an age of uncertainty caused by wars. The Vatican monument and the Dunaszekcsõ portrait are opposites of each other although still produced with the same idea: the Vatican monument could depict Marcus nakedly because the body was obviously not his, but Doryphoros', the Dunaszekcsõ portrait could present the emperor likewise as Serapis, because the work was nothing but a trustworthy portrait of Marcus Aurelius. It is possible to date this portrait to the years following the "Rain Miracle" because of the effect of the Alexandrian monument on it.

Adél Bugán
Aquincum Museum
Budapest

EGYPTIAN GODS ON THE EASTERN ADRIATIC COAST[1]

Haec tamen Aegyptia
quondam nunc
et sacra Romana sunt.
Minucius Felix

The Istrian peninsula was, from an administrative point of view, included in the Xth Italic region, under the name of *Venetia et Histria,* from around 16 BC,[2] while the immense *Illyricum*, which extended from the Adriatic to the Danube basin, was divided into the provinces *Pannonia* and *D a l m a t i a*[3] *1.). Dalmatia* occupied a comparatively large territory, with an administrative centre located in *Salona* (Solin near Split). The province began on the Adriatic at the river Raša in the north and at Istria in the west and extended as far south as Lissus (Lješ) at the mouth of the Albanian Drin (Drilo). In the 2nd and 1st century BC, howev-

[1] The area discussed is the coastline of the Republic of Croatia with the immediate hinterland.

[2] That is under Octavian, later known as Augustus, the only ruler of the Roman Empire after the battle of Actium in 31 BC

[3] The Roman *Dalmatia* included not only the Adriatic coastline, which still today bears that name, but also the Republic of Bosnia and Hercegovina and Montenegro, most of Southern Croatia, together with a large part of western Serbia. The border between Histria and Dalmatia was settled on the river Raša (*Arsia flumen*) in Western Istria. Dalmatia has a narrow coastline zone with rock-bound harbours and numerous islands, with an expressive Mediterranean climate and karst landscape with a little fertile land shut, mostly divided from the hinterland with mountain-chains.

er, to the Romans *Dalmatia* meant the area between the rivers Krka (*Titius flumen*) and Neretva (*Naro flumen*). It seems that this name already included, even before the establishment o the province, not only the territory of the Illyrian tribe of the *Dalmatae,* but also that of other similar tribes, such as the territory between the rivers Raša and Krka (*Titius flumen*), inhabited by the Illyrian tribe of the *Liburni,* which had a significant past and material culture.[4]

Isis was the dominant figure of the Egyptian cult sphere that spread through the Hellenistic and Roman world, but she was not the only Egyptian deity which the Hellens and the heterogeneous ethnic and social groups in the Roman Empire accepted. There were also Osiris and Serapis, Anubis, Horus or Harpokrates, Bes, the enigmatic figure of Jupiter-Amon, but they all have to thank their presence to their family- or mystery-ties to Isis. If we take a closer look at *Isis* or her ancient Egyptian form called *Iset*, they seem to be two different deities.[5] When the worship of Isis rose in Greek and Roman times only a small group of priests knew how to read the hieroglyphs. Thus the Greek and Roman method of superimposing their cultural values on other cultures easily affected the interpretation of the original *Iset*, turning her into a Roman deity. In her metamorphoses in the Hellenistic-Roman world Isis gained the characteristics of a supreme, omnipotent and universal divinity.

4 In the late prehistoric period, from the 9th century BC, and in historic times, the Liburnians inhabited the territory of northern Dalmatia between the river Krka (*Titius flumen*) and Zrmanja (*Tedanius flumen*), the islands, the coastal region below the Velebit mountains, the Croatian seaboard and the eastern part of Istria to the Raša river and the Quarnero islands. The Liburnians were certainly a separate cultural and ethnic groups.

5 PLUTARCH, *De Iside et Osiride*; FRASER 1967, 23-45; MÜNSTER 1968; VIDMAN 1970, 125-138; SELEM 1971, 269-293; VANDERLIP 1972, 1-106; GRENIER 1977, 169-175; CASTIGLIONE, 1978, 208-232; ASSMANN 1979, 93-115; BIANCHI 1980, 9-36; GRIFFITHS 1980; MORA 1990; DONADONI 1997, 32-36; MALAISE 1972; MALAISE 1972a; MALAISE 1978, 627-717; MALAISE 1984, 1629-1648.

HISTRIA

Colonia Pietas Iulia Pola was founded in around 45 BC, situated on the western coast of the peninsula. As an administrative unit a vast agricultural area (*ager*) belonged to the city, which was also a very important harbour on the Eastern Adriatic maritime route. The great achievements of Roman civilization are particularly visible in the architectural monuments and in town planning. The Roman colony of Pola retained a concentrically radial network of streets from the Histrian hillfort period (the so-called "spider nest" scheme) that was upgraded with planned architecture and with communal facilities. The lower southern part of the city by the sea was based on the scheme of regular orthogonal streets. The Forum was located here with two twin temples. That which was dedicated to *Augustus et Roma* is still preserved. The central temple was probably dedicated to the Capitoline triad. The city was adorned with numerous public buildings and with monumental private monuments (city walls with numerous quadrangular, circular and polygonal towers and monumental city gates, two theatres, an amphitheatre and luxurious private houses).

Solely in the city of Pola is there no direct evidence of the existence of a temple dedicated to any Egyptian divinity, but one of the findings is the beautiful marble fragment of a head depicting the goddess *Hathor* from the 2nd or 3rd century AD (Pl. 3/1).[6] She is represented with the characteristic cow-like ears and a headdress folded three times and tied with three gold plated bands. The nose is short and flat, the mouth disproportionately small and closed, with fatted lips and dimples. The chin is pointed, the slant eyes are unmarked, the pupils only engraved, the eyebrow lines cringed. There is a gilt band on a part of the neck, i.e. a necklace with triangular shaped incisions. The head was found near the south entrance of Pula's Amphitheatre. The whole relief application was attached to a vertical wall surface. There are several fixing holes on the rear, as well as the remains of mortar. There is a digit engraved on the rear: XIII. Similar figures of the goddess were found in Rome.[7] Near the walls of the Amphitheatre in Pola a tombstone with the bust of a deceased woman chiselled in a rectangular aedicule (Pl. 3/2) was also found.[8] The right side of the aedicule is flanked by a figure of a funerary Attis[9] in

6 White marble with golden plating, height 11.8 cm, width 17.4 cm, Archaeological Museum of Istria, inv. no. A 8712: JURKIĆ-GIRARDI 1976, 215; BUDISCHOVSKY 1977, 173, no. VIII/9.

7 ROULLET 1972, 56, XXII, fig. 33.

8 Limestone, height 92 cm, width 79 cm, thickness 36 cm, Archaeological Museum of Istria in Pola, inv. no. A 313 = CIL 5, 203; REICHEL 1983, 4-5, br. 84; I. I. X/1, 335; JURKIĆ-GIRARDI 1972, 49.

9 It is indicated by the position of the left hand, the preserved part of the legs,

low relief, freely sticking out. The left side is also flanked by a very badly damaged figure, possibly Genia. These types of figures were conceived as carriers of the gable, but also as symbolic escorts of the deceased. The outlines of two ithyfallic hermae can be seen on the sides of the stela in low relief. They have high hats on their heads, possibly of an Oriental *kalathos* type, or it could also be the hat worn by the Egyptian god Serapis. Under the aedicule the following text is to be found: *Obellia[e] Maximae*.

A beautiful and very unusual votive altar dedicated to Isis (Pl. 4 -5) originates from Galižana near Pola (*vicus Galisanus*).[10] It is made from a rectangular block of limestone places on a wider pedestal.[11]

The inscription is dedicated to Isis and some of the symbols and objects represented were used during the performances of her cult, but the relief decoration on the altar is not exclusively connected to her and her circle. If our interpretation of the symbols is exact, then it shows some singular syncretistic characteristics. On the front side (Pl. 4/2) we read the votive inscription: *[I]sids inperio / Q(uintus) Lutatius / Iucundus.* The following symbols are shown on the altar; front – a *sistrum*, a *situla*, a little prophet bird (*avis*), trident and dolphin (Neptune), *crotales* (Dionysiac circle), the bust of a cat (Isis-Bubastis), a barrel (Iupiter), one rython, one cup and one unrecognizable attribute (perhaps of Venus); left side (Pl. 5) – two crossed broad axes (*ascia*, connected to the cult of Isis, but also of Belona and Iupiter Dolichenus), crescent, *aulos* (double phrygian flute) or augur's crooked cane (*lituus*) and *cymbala* (Magna Mater), two

the rich clothing typical for this Phrygian god and it being positioned on a tombstone. The upper part of the body and the head is destroyed.

10 Many sites with architectural remains have been located in Galižana: MATIJAŠIĆ 1988, 39-40.

11 Height 91.5 cm, width 50 cm, depth 43 cm, Archaeological Museum of Istria, inv. no. A 41, ex no. 229 = CIL V, 510; I. I. X/I, 601; REICHEL 1983, 7-9, br. 91; WEISSHAUPL 1901; BUDISCHOVSKY 1977, 168-170.

amphorae, a syrinx (Pan), an oar and the tiller on a hemisphere (Isis Fortuna), stock with a wrapped snake (*caduceus* or *baculus*); right side (Pl. 4/3) – a crescent, a circle, a peacock above the earth, a lance (Minerva), pincers and a hammer (Volcano), a naked boy leading a dog (Harpokrates and Sirius), a sword (*gladius*), a vessel in the shape of an alabastron, a phallus, a palm-tree or ears (*spica*); and some indefinable symbols. The formula *Isidis inperio* determines the sole essence of the Hellenistic-Roman Isis as a universal, omnipotent goddess, united in a unique *imperium* with other divinities. I think, that the altar had to be exhibited in or around a shrine and it had to be visible from all three sides. It dates from the shift of the 1st to the 2nd century AD.

Two other votive inscriptions to the goddess Isis were also found in the Croatian territory of Istria. One comes from Poreč (***Colonia Iulia Parentium***): Isis Augusta,[12] the other from Novigrad (***Neapolis, Emonia***): *Isidi / Aug(ustae) sac(rum)... / v(otum) s(olvit) l(ibens) m(erito).*[13] On the south-eastern slopes of the Istrian Peninsula, at the tip called Vizače, remains of a prehistoric (centre of the Illyrian tribe of the Histrians), ancient and classical Roman town, ***Nesactium***, have been preserved through the centuries. In ***Nesactium*** the cult of the Egyptian goddess Isis is confirmed by a miniature bronze ritual rattle (*sistrum*, Pl. 6/1), discovered in 1908 in the necropolis in the vicinity of a wealthy family's tomb as a surface find.[14] It is made of an extended trapezoidal ring with four metal sticks. A cat or a dog is represented on the end of its handle. The *sistrum* was primarily connected to the cult of Hathor, rather than to other female divinities. With the expansion of the Egyptian cults in the Greco-Roman world, it became an attribute of Isis, her priests and her followers.

12 CIL V, 347; I.I. X, 2, 32; BUDISCHOVSKY 1977, 168.

13 MLAKAR 1957, 454-455; BUDISCHOVSKY 1977, 167.

14 Bronze, height 7.8 cm, lenght 5.2 cm, Archaeological Museum of Istria, Pula, invt. num. A 5048: PUSCHI 1914, 59-60; GNIRS 1915, 121; BUDISCHOVSKY 1977, 173, IX.

The appearance of the *sistrum* on frescoes, votive reliefs, altars, sepulchral monuments, sculptures and different objects of various characters and use leads us to create the opinion, that apart from its use as a ritual object, it must have also had a specific, probably magic, apotropaic role.[15]

Votive bronze statuettes must have been constituent parts of the Egyptian and Roman home-shrines of the inhabitants of Istria, but the sites from which these objects originate are with few exceptions unknown. A standing bronze figure of Isis-Fortuna from Savudrija (***Salvore, Silvium Promontorium***) represents her in her classical iconographical stature with the lowered right arm holding a tiller, while the left hand is holding the corn of abundance (*cornucopia*, Fig. 9).[16] It dates from the 2nd century AD. A statuette of Osiris from Bale (***Valle***) was discovered in a Roman tomb from the 1st-2nd century AD. It represents Osiris in the characteristic iconography of a mummified male figure with the crown of Upper Egypt on the head and the pharaonic attributes – the shepherd's crook *heka* and the whip *nekhad* – in the stylized hands. Unprovenanced items are the small bronze statuette of the sacred bull Apis with a disc between his horns, a bronze amulet-statuette with the representation of a lying naked man, a bronze statuette of a walking male figure and an Egyptian woman kneeling on a base.[17]

A group of shabti statuettes and amulets (scarabs and wedjat eye) are also kept in the Archaeological Museum of Istria

15 BALDACCI 1987, 239-244; GENAILLE 1976-1977; GENAILLE 1994; TIBILETTI 1997a, 660-661, 669-672, cat. XI. 9-16, bronze sistrums of the lenghts of 17.5 to 30.4 cm. Cf. For iconography - Plutarch, *De Iside et Osiride*, 63; its name as *crepitaculum* - Apuleius, *Metamorphosis*, XI 4 passim; *"children's rattle"* - Martial, *Epigr.*, XI, 53-54; Cleopatra shaked the *sistrum* during the battle of Actium, like Isis to incite the battle - Vergilius, *Aen.*, VIII, 696; ...

16 Height 9.2 cm, Archaeological Museum of Istria, Pola, inv. no. A 4620: DEGRASSI 1962, 881; BUDISCHOVSKY 1977, 167.

17 They are all kept in the Archaeological Museum of Istria in Pola, BUDISCHOVSKY 1977, 168, 174-175; JURKIĆ-GIRARDI – DŽIN - URANIĆ 2001, 18-21.

in Pola. They were part of the Austrian collection of an unknown donor, possibly from Phaestos on the island of Crete.[18]

DALMATIA

Following the coastline from north to south, we make the first stop in Senj (***Senia*).** It possessed an excellent harbour and was one of the few places where a road passed from the coast across the Velebit mountains into the Lika plain. The Roman city was established under Augustus and is listed among the *oppida* by Pliny, but it is not certain whether it was a *colonia* or a *municipium*..[19]

The fragment of a sculpture of Serapis with a votive inscription on the pedestal was discovered and described for the first time in 1898. By the time of its removal to the Lapidarium of the City Museum in Senj in 1962, the inscription had been completely erased. Owing to a former description, we know that *Sarmentius / Geminus / Sarapidi Deo Sancto* could be read in a *tabula ansata,* engraved in three lines.[20] In between, the lower part of a deity is represented *en face* in a seated position, wearing a richly draped long garment. The right leg is slightly stretched in front of him, while he is leaning against the throne with the left side of his body. Behind this Serapis, on the left, the remains of a column or part of a tree that he leaned on are visible. Behind his right leg, however, based on similar representations, the presence of an animal, Cerberus or lion, can be supposed. The sculpture is linked most probably to an oriental place, from where its donor, the libertine *Sarmentius* originated (Greece or the hellenised Orient). Based on the votive formula *Deo Sancto*, which is rare among the votive formulas of this Alexandrian god, the inscription dates at the beginning of the 3rd century AD.

18 URANIC 2001, 22-28.

19 Plinius, *Naturalis historia*, III, 140; GLAVIČIĆ 1993.

20 White marble, height 27 cm, pedestal 25 x 22 cm. BRUNŠMID 1898, 172-173; CIL 3, 15092; KLEMENC 1940, 6; SELEM 1972, 8; BUDISCHOVSKY 1977, 177; GLAVIČIĆ 1981-82, 66-68; GLAVIČIĆ 1994, 69-70; SELEM 1997, 47-48.

In the Lands Museum in Sarajevo (Bosnia-Herzegovina) a reddish cornelian gem is kept, supposedly originating from Senj. The gem represents the hairy and bearded head of a man with the *kalathos,* so it is obviously the representation of Serapis.[21]

Aenona (Nin) lies on the coast 18km north of *Iader.* It lies on a small island surrounded partly by the sea and partly by marshy lagoons, and is connected to the mainland by two causeways. In the pre-Roman period *Aenona* was already an important Liburnian centre. It became a complete part of the ancient world during the period of the consolidation of Roman rule in Dalmatia. Its municipal status was conferred under Augustus, apparently as early as 16 BC.

In 1852 a monumental statue of Isis with two inscriptions was found here (Pl. 6/3).[22] It is made from grossly grained white marble, which turned yellow on the surface. The head and neck are missing (they were made from a separate piece of stone, than fixed in the appropriate hole), as are the arms below the elbows. Owing to the general iconographical composition and the layout, we can recognise it as a representation of Isis. The goddess is standing erect with the left leg slightly curved backwards, while the right leg is firmly positioned on the pedestal, so the weight is on it. She is dressed in an ionic chiton, above which she wears a nicely draped cloak, tied in a knot above the bosom. The Isiac knot or the Egyptian *tjet* symbolised fertility and it represents the most typical Isiac attribute. Under the fine fold of the texture the luxurious bosom and the gently rounded legs are evident. Characteristically cringed locks of hair fall both on the left side of the bosom and on the right shoulder. The goddess held most probably a *sistrum* in her right hand, and a *situla* in the left.

21 BUDISCHOVSKY 1977, 176; SELEM 1997, 48. Measures of the gem 1.5 x 1.0 cm.

22 The sculpture has been kept in the Archaeological Museum in Zagreb. Height 136.5 cm, width 45 cm: KUBITSCHEK-LOEWY 1879, 166, no. 3; BRUNŠMID 1903-04, 230, num. 34; SELEM 1972, 10-11; BUDISCHOVSKY 1977, 176-177; SELEM 1997, 54-55, no. 1.11.

The iconographical model of this Isis belongs to the Greek art of the second half of the 4th century BC. Images of goddesses and mortal women were not formally always fully distinguished, and in many cases it is hard for us to tell them apart. In this case, however, it can be excluded that this is the representation of a priestess of Isis, for it is not even remotely probable that in *Aenona* at the end of the 1st century AD such a monumental statue would have been erected for her priestess. This statue belonged, most probably, to a temple or shrine of the goddess (*iseum*), the position of which is, however, questionable.
Furthermore, two lamps with the busts of Isis and Serapis one lamp with the representation of Harpokrates and one with Jupiter-Amon were also found in *Aenona*. Both lamps with the busts of Isis and Serapis have the representation of the confronted busts of the divinities in the central medallion (Pl. 7/1 and 2).[23] Isis is represented on the left, with the lotus flower on the head, the hair collected in a knot. Serapis is shown on the right, with a rich beard, the *kalathos* and the rayed crown. The two busts together appeared for the first time in Alexandria on staters of Ptolemy IV Philopator, around the year 217 BC.[24] The motif of confronted busts seems to be from the Roman period. It appears on a series of lamps produced in Italy. It is connected to the ascension of the Egyptian cult there, most stated during the Flavian Emperors. The motif, in particular variations, appears again in the second half of the 2nd century, on a series of terracotta lamps, produced in *Africa proconsularis.* Most of them come from Tunis, and appear on a series of Gaul-Roman medals produced in the valley of the Rhône. Our lamps belong to the

23 First lamp, length 10.1 cm, width 7.3 cm, thickness to 4 cm, found in Nin in 1896, it is kept in the Archaeological Museum in Zadar, inv. no. L. 163, I.e. 6820, t.k. 4379: BERSA 1906, 53, no. 181, tab. 3, 1; SELEM 1997, 55-57, no. 1.13. Second lamp, lenght 8.8 cm, width 7.7 cm, thickness 3.4 cm, discovered in Nin, bought in 1924 for the Archaeological Museum in Zadar, inv. no. L. 878, I.a. 1383/2, t.k. 4427: SELEM 1997, 57-58, no. 1.14.
24 TRAN TAM TINH 1970.

Tunis types (Loeschke VIII). The symbolism of the motif is similar to that on coins and commemorative medals with the representation of the Imperial couple. Such representations of Tiberius and Livia, Nero and Octavia, Commodius and Crispina, Alexander Severus and Iulia Mammea are known. When, during the 2nd century, the Alexandrian couple of gods came in close connection to the Imperial cult, the representations of Serapis and Isis symbolised the divine harmony at the leadership of the Empire.

The lamp with the standard iconographical model of Harpokrates is of a simple type with rounded beak 13).[25] A naked chubby boy with the head turned to the right stands on the slightly convex front side in high relief. The finger of the right hand is brought to the mouth, while in his left hand he holds a lotus flower. On the head he has the stylised *psent* crown. The central medallion of the lamp with the representation of Jupiter-Amon bears the head of a hairy god with ram's horns.[26]

It was discovered in Nin, in a necropolis of the early Roman period. Two other interesting lamps come from Nin. One, in the central medallion, has a sculpturally represented crocodile, the sacred animal to Isis-Thermutis,[27] turned to the left, while

25 Brownish terracotta, height 12 cm, it is kept in the Archaeological Museum in Zadar, inv. no. L. 596, I.a. 706, I. e. 6771, t. k. 4990: BERSA 1915, 59, num. 676; BUDISCHOVSKY 1977, 181; SELEM 1997, 55, num. 1.12.

26 It was kept in the Archaeological Museum in Zadar, but now impossible to trace: ABRAMIĆ 1912, 101; BUDISCHOVSKY 1977, 177; SELEM 1997, 58, no. 1.15.

27 The assimilation of Isis and Renenutet, i. e. Hermutis or Thermutis began in the Ptolemaic period in Narmouthis, today's Medined Madi. She was celebrated e.g. by the Greek-Egyptian poet and priest *Isidorus* in the 1st century BC In one of the iconographical varieties, Isis is represented nursing a little crocodile, Sobek-Horus. The syncretism between Isis and Neith, mother of the crocodile Sobek, celebrated in the Faiyum, can be noted. The lamp from Nin represents the baby Horus riding a crocodile., cf. BRESCIANI 1997, 37-41.

three stylized flowers and leaves can be seen in the background.[28] The second lamp presents a half-naked man riding a crocodile.[29] Around the flanks he has an apron and in each hand he holds a double flute.

Iader (Zadar) became a Roman colony at the end of the 1st century BC, a city second only to the provincial capital *Salona*. The Roman city was built on the peninsula on which even today the centre is located. There is a good harbour between the peninsula and the mainland. Before the Roman period the place was an important Liburnian settlement, attested to by numerous finds of the pre-Roman period. The town was divided into symmetrical blocks surrounded by a rectangular raster of streets, according to the Roman town-building rules. The centre of the town, the Forum, is the most spacious of all forums explored in Croatia (it measures 95 ms in length and 44.5 ms in width). Its origin dates from the time of the foundation of the colony and it assumed its final aspect after reconstruction carried out in the 3rd century AD. In the western extension of the Forum there is a monumental Capitol, in the middle of which was erected the temple. The temple (*prostylos hexastylos*) had three *cellae* and was dedicated to the Capitolian triad (*Juppiter, Juno, Minerva*). A portico (peribolus) framed the whole space of the Capitol.

By shape and dimension, the most monumental votive altar discovered in Zadar was dedicated to Isis and Serapis, Liber and Libera.[30] Its tall base was separated from the central part by

28 Reddish terracotta, lenght 8.07 cm, width 6.06 cm, height 2.1 cm. Found in Nin, bought for the Archaeological Museum in Zadar in 1908, inv. no. L. 702, I.a. 584/6, I. e. 7388: BERSA 1915, 56, no. 685.

29 Reddish terracotta, lenght 11.7 cm, width 8.3 cm, height 3.0 cm, found in Nin in 1897, it is kept in the Archaeological Museum in Zadar, inv. no. L. 381, I.e. 7247, t.k. 4423: BERSA 1915, 67, num. 241.

30 Limestone, height 94 cm, width 59 cm, depth 46 cm, Museo archeologico Maffeiano - Verona, inv. no. 246: CIL 3, 2903; BUDISCHOVSKY 1977, 178-179; MODONESI 1995, 59-60, no. 58; GIUNIO 1997, 269-270, cat. 14.K; SELEM 1997, 49-50, no. 1.4.

a simple profilation (it is lacking on the left lateral side due to greater damage on this part), which we find also on the top of the altar. On the crown of the altar are positioned the two rounded torqued cushionets as *acroteria*, tied in the middle with a ribbon. Inside the cushionets is a small relief rosette. On the front side there is a triply profilated inscription area (29 x 39 cm) (Pl. 8/1). The letters are quite regular and nicely engraved and there are also traces of the three-cornered punctuation marks (height of the letters 3.8-4.3 cm). The inscription reads: *Isidi Serapi(di) Liber(o) / Liberae voto / suscepto pro salute / Scapulae filii sui / P(ublius) Quinctius Paris / s(olvit) l(ibens) m(erito).*

On the left side of the altar we can see the standing figures of Isis and Serapis, dressed in a tunica covered up with a cloak (Pl. 8/2). The faces are quite badly damaged. Isis's hair falls in locks on the left shoulder. She is holding a long sceptre in her left hand, as is Serapis, and a *sistrum* in her elevated right hand. On the right side of the altar, in a panel of equally simple profilation, two sitting figures are represented, male and female, most probably Liber and Libera (Pl. 8/3). On the rear, inside a deeper field and on the sides of a column, two male standing figures are represented (Pl. 8/4). Most probably these are the figures of Anubis and Harpokrates, although their identification is aggravated by the damaged surface.

The altar found in Zadar *prope aedem S. Salvatoris* is cited in the *CIL*, but the year of the discovery of this monument and how it arrived in Verona are still unknown. We know that Lucius cited the inscription in the year 1673 and that Maffei mentions it in his catalogue of the museum in Verona in 1749.[31]

Scipione Maffei was a private collector from Verona, who attained his collection mostly by exchange, purchase or gifts. His

31 LUCIUS 1749, 84, 3. MAFFEI 1749. We know that in the year 1623 the medieval church of St. Savior was partioned and was than named St. Anthony the Abbot. It is possible that our altar was discovered at that time. During the Second World War the church was completely destroyed.

collection afterward appertained to the city and is today the nucleus of the Museo Archeologico Maffeiano. The quotations about the location of where the altar was found, all indicate the vicinity of the city harbour, and Vitruvius in his chapter about the choice of building ground for public buildings states that the temples of Isis and Serapis should be raised in harbours.[32] The altar was raised by the vow taken (*voto suscepto*) by libertine *Publius Quinctius Paris* for the health of his son *Scapula*. Although, it was stated, the altar couldn't have been raised before the first half of the 2nd century, in my opinion its raising could be stretched into the late 1st century AD.

During the archaeological excavations of the Capitolian Temple in Zadar in 1964, among the foundation walls in the *favissae* of the Temple, a quadrilateral press-model with the negative representation of Serapis was discovered (Pl. 9/1).[33] The frame of the composition represents the front side of an *aedicula* (*naiskos*) with columns on the side, with capitals and the gable. The press-model is quite threadbare and some parts are damaged, but all the major details can still be seen. Primarily (first brought up by M. Suić and accepted by P. Selem) the central figure was attributed to the Egyptian god Serapis, whose iconography was obviously enriched with numerous sincretistic elements. This attribution was, however, denied by N. Cambi and J. Medini, who were the first to write about it in detail. The analysis of all the iconographical elements brings up rather the identification of the Phrygian divinity Sabazius. As in most places the figure is still characterized as Serapis, I have enlisted this monument in my article.

The frame of the *aedicula* is made up of two columns with capitals fastened with a torqued horizontal beam, which is the base of the gable. In the centre of the gable an eagle

32 Vitruvius, *De architectura*, I, 7: "*... ut Isidi et Serapi, in emporio*".

33 Sandstone, dimension 17.5 x 10 x 5 cm, Archaeological Museum in Zadar. SUIĆ 1965, 97-100; CAMBI 1963-65, 103, note 102; SELEM 1972, 9, 54-58; BUDISCHOVSKY 1977, 180; MEDINI 1980, 67-88; SELEM 1997, 50-51, no. 1.6.

with opened wings and thunderbolts in his claws (an iconographic attribute of Iupiter) is represented, while between the columns the standing figure of a mature man with luxurious hair and beard can be seen. He is dressed in a very short oriental chiton, probably with long sleeves. He has a cloak thrown over his shoulders and buttoned around the neck. On his head he is wearing the cylindrical mystique basket, the *kalathos*, unusually short because of the vicinity of the gable beam. The divinity is represented *en face* leaning on his left leg, while with his right leg he takes a step forward on a ram's head. In his right hand he holds a pinecone and with his left one he reaches for a tree with an entwined snake. The snake's head is directed towards the god's head. Suić identified it as the press-model of an amulet with a figure holding the elementary attributes of Serapis. But many other attributes are present here from the iconographies of other oriental and some Greco-Roman divinities (for example the snake for Esculape, the ram's head for Jupiter-Amon). The oriental suite has special iconographical value. Serapis does not wear such clothes. He took over his clothing in the Hellenistic period in sincretism with Jupiter. The pinecone held in the right hand and the tree reached for by his left hand have been taken as the iconography of Sabazius. The posture of the figure and his oriental clothing agree to an extraordinary extent with the representations of Jupiter Sabazius on bronze plaques. There are not many analogies, in fact only three: two from Rome (one kept today in Berlin, the other in Copenhagen) and one from Ampurias in Spain.[34] On these plaques the divinity is represented in the same posture as our Sabazius. He wears the same oriental clothes and tight trousers. Trousers were most probably also present in the iconography of the divinity from Zadar, but the state of preservation of the press-model from Zadar does not allow us to distinguish such details. On all the plaques Sabazius is represented with his right leg taking a step

34 BLINKENBERG 1904, 91-, pl. II; BELLIDO 1967, 73, fig. 5 a-b.

forward onto a ram's head. Thus this press-model represents the sincretism of the oriental divinities Sabazius and Serapis with Iupiter. It must have been used for producing metal, bronze or even silver plaques for the cult in the underground rooms of the Temple of Capitolian divinities or by the very temple under the north-eastern peribolos where underground rooms were discovered during archaeological excavations. It can be dated to the 2nd or the beginning of the 3rd century AD.

During archaeological excavations in Zadar in 1990, an amulet made of lead-bronze was discovered, made according to the iconographical scheme of the mummified figure.[35] It is quite badly damaged. The base and the sting for fixing it into the base (usually the sting sprouts from the feet of the figure) are missing. The amulet represents the Egyptian God Osiris.[36] Originally Osiris was represented as the king, dressed in a tight cloak and white skirt and as a god of the underground he acquired the shape of a mummy. The Osiris from Zadar wears the pharaonic wig on his head. It is also possible that he had the characteristic necklace *usekh*. The nose, the eyes and the oval face are still recognizable. Above the forehead, in line with the mouth and the chin, a small hole is visible.[37] The hands are free from the funerary mantel and are crossed on the chest, the palms closed, posed one above the other.[38]

35 Max. heigth 3.7 cm, width 1.05-1.8-0.8 cm, thickness 1.2-1.5-0.8 cm. Weight 33,3 grams. GIUNIO 2000, 177-191.

36 At the time of the finding and at the beginning of my studies, the shape of the amulet lead me to the supposition it was a *shabti* figure, the Egyptian funerary statuette. During my researches I found out the unexistance of *shabti* figures made of bronze: SPEELERS 1923; PETRIE 1935; DONATELLI 1988, 198-211; LISE 1988, 131-133.

37 It is possible that here stood originaly the *uraeus,* the royal cobra, symbol of light and royal dignity. It was sacred to the goddess *Uadjet* (*Uto*, the goddess protector of the Lower Egypt) and to god *Ra*.

38 The explanation of the position of the hands: ROEDER 1937, 89-96, 346, pl. 12 a-b.

He is holding the *flagellum* i. e. the *nekhekh* (the leather interwoven whip with three fringes, the symbol of power) on the right shoulder and the sceptre *heka* on the left shoulder (in the shape of the shepherd's crook, symbol of the pharaonic authority). This Osiris lacks the *atef* crown on the head and the royal beard.[39] It is a product of local provincial workshops, although the execution is iconographically faithful to the original Egyptian type because of the rustic elaboration and the dumpy look of a comparatively short body. It can be dated to the period between the 1st and the 3rd century AD.

In the Lands Museum in Sarajevo, a gem made of black onyx represents a bearded man with the *kalathos* on the head, therefore Serapis. It is supposed to have come from Zadar.[40]

Varvaria is a Liburnian hill-settlement with defences of the pre-Roman and Roman period on Gradina near Bribir. The *Varvarini* were granted tribute exemption early under Augustus and they probably acquired a municipal character under Tiberius. In 1977 a small votive altar with a dedication to Isis was discovered here: *Isidi / sac(rum) / Raecia Mar/cella* (Pl. 9/2).[41] The altar dedicated to Isis is one of the many small dimension altars (*ara – arula*) discovered on the territory of the ancient *Varvaria*. Its shape is typical, made in the spirit of the local mentality for featuring votive monuments. They are of a small size, made of a profilated base, the inscription area and the profilated head with volutes. The altar of Isis dates from the first half of the 2nd century AD.

39 Cf. Osiris without beard: statue of Osiris from the Louvre, Paris; statue of Osiris from the Roman period, Musei Vaticani, Roma. Osiris without beard and crown: the statue of Osiris from the Museo Nazionale Romano, inv. num. 58206; statue of Osiris from the Museo archeologico, Torino. Osiris with beard and *uraeus*: the lamp in the shape of Osiris from the Museo Archeologico Civico "U. Formentini", La Spezia, inv. no. 1072.

40 BUDISCHOVSKY 1977, 181; SELEM 1997, 54, num. 1.10.

41 Limestone, dimension 29.5 x 19 x 13 cm. It is kept in the Lapidarium at the site: KUNTIĆ-MAKVIĆ 1982, 151-157.

SALONA. **S**ix kilometres north of Split, situated at the heart of present-day Solin is the archaeological site containing ruins of *Salona*, the principal town of the Roman province of Dalmatia. A favourable central geographical position on the eastern coast of the Adriatic at the bottom of the well-sheltered Bay of Kaštela, along the delta of the river Salon (today the river Jadro), as well as good road links to the hinterland all contributed to the quick and unhindered development of the town. Salona was settled in as early as the prehistoric times, but it was only under Roman sway that it developed into a large town with over 60,000 inhabitants. The colonia was established with the titles *Colonia Martia Iulia Salona* at some point between 47 and 27 BC.

Initially, Salona was the coastal stronghold and the port of the Illyrian Dalmats in the immediate vicinity of the ancient Greek colonies Tragurion and Epetion. Along with the local Illyrian population and the Greek settlers, Salona was at the time inhabited by a large Italic community. Following the civil war between Caesar and Pompey in 48 BC, Salona was first granted the status of a *conventus civium Romanorum*, then of a Roman colony thus becoming the centre of *Illyricum* and later of the province of *Dalmatia*. After the last Illyrian rebellion had been quelled (Baton's Uprising, 6th-9th years AD), Salona entered a new period of peace and prosperity, as illustrated by its urban development and intensive construction activity.

A shrine or temple dedicated to the goddess Isis certainly existed here, as a statue of the goddess in natural size is described and other evidence of her cult was found there.[42] Of primary importance are the three bronze busts of Isis, Serapis and one of an anonymous winged male figure, discovered during the first

42 Lost today, there is a fragment of an inscription with the name of Isis from Salona: CIL 3, suppl. 2, 12837; SELEM 1959, 96, note 5.

43 Kunsthistorisches Museum Wien, inv. no. VI 395: SACKEN 1871, 90, no. 6 and 7; DREXLER 1900, 146-148; BUDISCHOVSKY 1977, 198, no. 7 and 8; SELEM 1972, 15; SELEM 1997, 70-73. no. 2.5 and 2.6.

archaeological excavations in 1820-1821.[43] The bust of Isis (height 18 cm) presents her in the image of the Roman goddess of fertility (Pl. 10/1). The goddess wears a dress tight to the body and opened in an irregular V-cut on the front. Above the dress the typical Isiac scarf with fringes is thrown over. They are nicely visible on the right side of the goddess's bosom and in the middle it is joined in the great Isiac knot, which extends from the neck's cut to the lower part of the bust. The head is slightly declined to the left. The hair is collected in a garland on the top of the head, and on the back it becomes a tail. Above the garland the front is framed with curls which are transformed on the sides into long locks. On her head she is carrying the royal diadem entwined with two large snakes. The snakes should represent the transformed cow-like horns, so in fact they would stay as Hathor symbols.[44] The singularity of this bust reveals that it is not an abstract face of a divinity, but that it bears some portrait features. It dates from the late Imperial period, perhaps the 3rd century AD.

The second bust (height 15 cm) represents Serapis, with a luxurious beard and hair that falls upon the back in long and slightly interlaced locks and overshadows the face (Pl. 10/2). On the head he has the *kalathos*, entwined with ribbons on each side and decorated with an olive branch, with one ear on each side. Serapis is wearing a garment with no sleeves but opened in a V-cut on the front above the neck and with a cloak thrown over the left shoulder. The Salonitan Serapis is a solid example of the mature art of the Imperial period, which in the Roman way transmits the model of the Hellenistic protector god.

The third figure represents a kneeling young man dressed in a short Egyptian apron with two opened wings on his back. The hands descend to the sides of the body and the palms are closed. His Egyptianized style is accentuated, but he doesn't bear any indication of his connection to the cult. It may have just a decorative purpose. His finding together with the busts of Isis and Serapis indicates, however, that he could have been a part of an Egyptian cult entity.

44 Isis-Hathor=Isis-Thermutis, see note 25.

Also the altar discovered in Pescara (Italy, Roman *Aeternum*) speaks of the cult of Serapis in Salona. It was erected for the mariner *Lucius Cassius Hermodorus*, a member of the *collegium Serapis Salonitano*.[45] Among the manuscripts kept in the Archaeological Museum in Split, a Latin document has been discovered which tells us about a very important archaeological find in Salona. It is titled - "*Cupreum Salonitanae Iunonis simulacrum illustratur*".[46] The notes at the bottom of the pages suggest that the document was written with the intention of being published. According to the paper and the handwriting it can be dated to the shift from the 18th to the 19th century. The anonymous writer calls this bronze statue the statue of Juno, basing his judgement on the appearance of the statue, mostly the gesture of lactation. He describes the goddess as full of the mature feminine dignity, dressed in a woman's stola which falls down to the feet, with a scarf decorated with fringes harmoniously combed in three lines and falling in front on both shoulders. The right hand is lowered to the knee, while the left hand is brought up toward the right breast, where the nipple is shown, as if she wants to squeeze the milk. Later on it is cited that the head of the sculpture is adorned with a radiant crown, which presents the greatest difficulty for interpretattion.[47]

Considering the description, as the sculpture has been lost for us, M. Nikolanci interpreted this sculpture as most probably being that of the Egyptian goddess Isis. Owing to the writer's

45 CIL 9, 3337; DREXLER 1900, 145-146; SELEM 1972, 14-15; BUDISCHOVSKY 1977, 28-29, no. II, 1; SELEM 1997, 62-64, num. 2.2.

46 NIKOLANCI 1963-65, 119-126; BUDISCHOVSKY 1977, 194-198, no. 6; SELEM 1979, 84-88; SELEM 1997, 67-70, num. 2.4.

47 "*Simulacrum est ex cupro conflatum, matronali maiestate praestans, muliebrique stola ad imos usque pedes demissa indutum: pectorali zona laciniis exornata praecingitur: trisulcis crinibus ad venustatem compositis, atque super utrumque humerum mammilari velo protuberantem non secus ac lac ex pectore expressurum: radianti demum corona simulacri redimitum est caput, quod plurimum negotii consideranti facessit*".

description of the scarf thrown over the goddess's shoulders, the attribution to Isis is possible. Although the scarf is not tied in a knot on the front, the fringes exclude our doubts. In the Hellenistic-Roman period the scarf with fringes is a determinant for Isis. This sculpture could be enlisted in the category of the Hellenized representations of Isis, where the Egyptian attributes are contaminated with the attribute of Greco-Roman divinities, in this specific example with Hero-Juno. A special riddle is her representation with a radiant crown, which is not a typical Isiac attribute. We can ask ourselves – not without reason - whether it is really a crown with rays or whether Isis was wearing many complicated cap-horns, the crescent, a plume and ears, which the anonymous writer – because of his ignorance – had described wrongly. If it was really a crown with rays, than it could be a substitution for the ancient sun-disc. The usual cap of her divine husband Serapis is the *kalathos*, but during the Roman Empire, especially from the 2nd century AD, a strong assimilation of Serapis with Sol, i. e. Helios begins. Serapis with the radiant crown became a common iconographical form at this time. Also Harpokrates, the third participant of the Alexandrian divine triad, can also wear the same crown.[48]

The anonymous writer also describes the pedestal of the statue decorated with pomegranate flowers and shields.[49] The decoration on the circular pedestal connects Isis to the army, i. e. to the Roman Imperial cult.[50] He doesn't give us the statue's dimensions, but from the description we can presume that it was nearly of natural size. The author only mentions that the statue

48 VIDMAN 1965, 390-394.

49 *"De basi etiam, cum Iunoni nostrae simulacrum insidet, uno verbo. Est basi ipsa orbiculatae figure, scutis, floribusque ornata".*

50 The known connection between Egyptian gods and the Roman army is the figure of the so-called military Horus, where he is represented anthropomorphically but with a falcon head, dressed a Roman legionary. Anubis has also been represented as a centurio: BENEDITE 1904, 115; PERDRIZET 1921, 35, tab. XXXII; ERMAN 1952, fig. 140; WESSETZKY 1961, tab. II, fig. 4.

was found in the ruins of ancient Salona, probably near the site where the remains of a now unknown ancient temple were. The model of this Isis is obviously Alexandrian; she is a dignified matron seated on a throne with a gesture of the act of lactation – *Isis lactans*. Following the trace of *scutum* and the possible connection with the Imperial cult, the statue can be dated to around the 2nd century. The appearance of Egyptian gods in Roman military uniforms (Horus, Anubis) or with military symbols – the lance and the shield – becomes intensive at the time of the Emperors Domitian and Traian. At the time of Emperor Commodus, Isis and Serapis became the protectors of the Imperial person. The statue was certainly honoured in a public place, so the connection with the complex Imperial cult is very probable.

Another testimony to the cult of Isis is the cover of the sarcophagus of the boy Aurelius Satrius and his sister Aurelia Maxima, discovered in 1880 on the site Smiljanovac, in the vicinity of the road *Salona - Epetium*.[51] It is the cover of a sarcophagus that was hollowed out of rock (Pl. 11/1 and 2). The cover dates from the third or fourth decade of the 4th century AD. Two medallions are represented on the covers. The left one shows the figure of Aurelius Satrius dressed in a wide white dress (*palla contabulata*) with the so-called Horus's lock on his head (Pl. 11/3), which designates the members of Isis's religious community in the Roman iconography of the 2nd century.[52] The diptych beside his head can be explained as the proof of his being instructed into mysteries and the religious

51 Dim. 234 x 129 x 63 cm, Archaeological Museum in Split, inv. no. A 393 or D 86: CIL 3, 8921; ABRAMIĆ 1924-25, 68-70; CAMBI 1963-65, 85-97, no. 1; BU-DISCHOVSKY 1977, 192-193, no. 3; CAMBI 2000, 88, note 633, fig. 144.

52 GONZENBACH 1957. She explained the origin of the lock as an African-Hamite (now afro-asiatic) custom which had the purpose of signing up a group of boys of a certain age, who were instructed collectively before they entered in the tribal community. From this cultural circle the lock was than moved Horus, and via him to the Isiac religion. Symbolically the meaning of this lock in the cult of Isis is signing the underaged, not yet full powered members of the Isiac religious community.

doctrine, i. e. the proof of his status as neophyte. The right medallion represents a little girl, Aurelia Maxima, on whose head there are two ribbons – one surrounding the head, the second axially over the top of the head (Pl. 11/4). From the crossover of the two ribbons hangs a little ball as a decorative pendant. In her hand she holds a bird. The signs are less evident, but they are also connected to the Isiac iconography. The flowery motifs on the sarcophagus are interpreted as the lotus pods.

The inscription was carved in the central medallion and also extended to the border band of the cover on the front and partly on the right side. There is a segment of the inscription under the figure of the little girl (a group of letters) which gives us some problems. The members of the *collegium isiacum* of Salona were buried in this tomb, so the solution to the mysterious abbreviations in the inscription is possible to find inside the Isiac religious concepts. The inscription should be read as: *D(is) M(anibus) / Aur(elio) Satrio / qui vixit a/nn(os) VIII m(enses) IIII / et Aur(eliae) Max/imae quae vi/xit ann(os) VI m(enses) III fili(i)s piissimis Aur(elius) Flav(u)s et Cer(e)s par(entes) et Is(idis) Vic(tricis) Vr(aniae ?) (et) Sa(rapidis) Con(servatoris) Alum(n)is P(ublius) Hipetius et ...* It is incomplete, the second name and ***posuerunt*** are missing. *Isis Urania* is not known from elsewhere, but Isis often had the symbolic meaning of the goddess of Heaven, and in the Roman period she was often called *Regina coeli* or *coelestis*. Furthermore, Isis in the *Aretalogy* from Chios was mentioned as the daughter of Uranius, i. e. Kronos.[53] The inscription should be explained as the dedication of two parents, Aurelius Flavius and Ceres, to their prematurely deceased children; Aurelius Satrius died at the age of eight and Aurelia Maxima died at the age of six. The inscription is dedicated at the same time to Isis, the heavenly Winner and Serapis the Saviour and moreover the dedicants were joined by Publius Hipetius and another person, whose name should come at the end of the inscription after the conjunction *et*. He

53 FESTUGIERE 1972, 138-163.

taught the children faith with Hipetius, as the term *alumni* was used for the students in faith. Publius Hipetius was one of the *mystagogos*, or the religious teacher, in the community.

As evidence for the Egyptian cults in Salona there are also two reliefs featuring a Nilotic landscape. Most probably they belonged to a fountain decoration or the crown of a well.[54] Taking in consideration the existence of single Isis' or Serapis' temples in other Roman cities, for instance Rome and Pompei, the well might have been situated in the enclosure of a Salonitan temple.[55]

In addition to that, a great number of statuettes of the Egyptian divine couple Isis and Osiris and other finds were found in Salona which confirm the existence of a huge Egyptian ethnic group in the town. Five bronze statuettes have been discovered representing Isis-Hathor nursing Horus,[56] the well-known type of *Isis lactans* (Pl. 12/1 and 2).[57] The goddess is dressed in a long and tight dress, the head adorned with a huge cap with cow-horns, the sundisc and the *ureus*. She is holding the little Horus with her left hand, while her right hand is brought up to her left breast. The feet are leaning on a shallow pedestal supplied with a thorn for fixing the statuette in her pedestal. A lock of Horus' hair falls toward the right ear (*Horuslocke*). Also six bronze statuettes of Osiris were found here (Pl. 12/3 and 4).[58] He is represented in the form of a mummy, the body wrapped with funeral

54 PRIJATELJ 1954, 29-35; BUDISCHOVSKY 1977, 199, no. 9, 10.

55 WILD 1981; HOFFMAN 1993; GALLO 1997; ...

56 Height of the statuettes 8.1-8.3-9.9 27 a and b)-10.0-11.5 cm, Arhaeological Museum in Split, inv. no. H 5003, 5004, 5005, 5007, without no: DREXLER 1900, 146; SELEM 1959, 98-100, no. 1-5; SELEM 1971, 17; BUDISCHOVSKY 1977, 199-201, num. 11-15; SELEM 1997, 73-74, no. 2.7-2.11.

57 TRAN TAM TINH 1973; LECLANT 1976, 89-101, fig. 1-12.

58 Height of the statuettes 6.628)-7.029)-8.0-8.9-9.8-11.7 cm, Archaeological Museum in Split, inv. no. H 374, 375, 376, 377, 378, 5006: DREXLER 1900, 146; SELEM 1959, 100-104; SELEM 1972, 17-18; BUDISCHOVSKY 1977, 201-203, num. 16-21; SELEM 1997, 75-76, no. 2.12-2.17.

textile, the head topped by the *atef* crown. In his hands, which are crossed on his chest, he is holding the *flagellum* and the *heka* sceptre. In Salona, respectively in the Salonitan territory, a small silver pendant was also discovered. It is in the form of the jackal god Anubis with the caduceus and the palm-tree.[59] Another find is a smaller, blue amulet-pendant representing the god Bes in his typical grotesque form which is an imported quality work of handicraft of the Roman period.[60] Furthermore, the bronze statuette reproducing Hermes-Thoth should be mentioned (Pl. 12/5),[61] and applique or phalera or the cover of a vase with the figure of Jupiter-Amon (Pl. 10/3),[62] and a phallic pendant-amulet, the type of which is generally connected to the cult of Osiris.[63]

A lamp was also found in Salona with the representation of Isis and Serapis and a lamp with the representation of the young god Harpokrates. On the first lamp, in the central medallion with the profilated radiant ornament on the exterior side of the border, two figures are represented.[64] A seated male figure in an ample garment occupies the central position. His cloth is richly draped and his right hand lies on the head of a lion situated at his feet. The *kalathos* identifies him as Serapis. In the left hand he holds an object similar to a shepherd's crook. On his left side stands a woman with her face framed by a veil, in a long draped dress, the right hand brought up to the breast. The gesture can symbolically

59 Transfered to Venice, lost today: DREXLER 1900, 149; SELEM 1972, 18; BUDISCHOVSKY 1977, 204, no. 23; SELEM 1997, 76-77, no. 2.18.

60 H. 2.8 cm, kept in the Archaeological Museum in Split, inv. no. G 349: LANZA 1856, no. 4; SELEM 1959, 105-106, no. 18; BUDISCHOVSKY 1977, 210, no. 39; SELEM 1997, 90-91, no. 2.53.

61 H. 6.7 cm, Archaeological museum in Split, inv. no. H 4081: ŽANIĆ-PROTIĆ 1988, 26-27, no. 17; SELEM 1997, 96, no. 2.61.

62 Bronze, diam. 3.9 cm, thickness 0.9 cm, Archaeological museum in Split, inv. no. H 3875: CAMBI 1980; SELEM 1997, 96-97, no. 2.62.

63 H. 3 cm, Archaeological museum Split, inv. no. 1221: SELEM 1997, 97, no. 2.65.

64 Archaeological museum Split, diam. 8 cm: BULIĆ 1887, 192, no. 443; BUDISCHOVSKY 1977, 211, no. 43; SELEM 1997, 77, no. 2.19.

be the sign of *Isis lactans.* In the left hand she is holding an object which most probably represents the corn of abundance with the *pyramidion* on the top. The lamp dates from the 1st century AD. In the second case a square base, decorated on the front side with vegetable motifs, features two lamps above two identical, standing chubby boys. This is Harpokrates according to the attributes (Fig. 11/6):[65] The finger of the right hand is brought up to the mouth, in the left hand there is the corn of abundance. On the head he wears a sign which can be explained as the stylized crown *psent* or the lotus flower.

Five gems with the representation of Serapis were also discovered in Salona,[66] three gems with the representation of Harpokrates,[67] and two representing Anubis (Fig. 9/4).[68] The Egyptian magical or gnostic gem-amulets bring up some yet unexplained subjects of sincretism of the Hellenistic-Roman period.[69] The appearance of various amulets is connected to the strengthening prophylactic and magical character of the divinities of the Isiac and Osiriac circle in later periods.

All these findings refer us to the conclusion that a devel-

65 Pink terracotta, height 6 cm, diam. 10 cm, Archaeological museum Split: BULIĆ 1886, 127, no. 1-2; BUDISCHOVSKY 1977, 211, no. 44; SELEM 1997, 77, no. 2.20.

66 The gems present Serapis's bust with luxurious hair and beard and with the *kalathos*. Particular is the gem representing a male's head with ram's horns and the *kalathos*. Above there is the eagle, on the sides of two unrecognisable objects, possibly military signs. It's a sincretistic figure of Serapis Amon Jupiter. BUDISCHOVSKY 1977, 211-212, no. 45-47; SELEM 1997, 78, no. 2.21-2.25.

67 Harpokrates is represented as a young boy, on the head he wears a crown with the lotus flower, in the right hand he holds the horn of abundance. BUDISCHOVSKY 1977, 212, no. 48; SELEM 1997, 79, no. 2.26-2.28.

68 Anubis as Hermanubis, represented as the figure of a naked man with the jackal's head and the Egyptian apron around the waste. In the stretched left hand he holds a bird with opened wings, in the right hand the *kerykeion*, i. e. the *caduceus*. BUDISCHOVSKY 1977, 212-213, no. 49; SELEM 1997, 79-80, no. 2.29-2.30.

69 SENA CHIESA 1997, 151-159.

oped Egyptian, i.e. Isiac, religious community and by all means a temple or shrine existed in Salona. This community lasted until the beginning of the 4th century imposing the systematic education on the youth. The examples are the deceased Aurelius Satrius and Aurelia Maxima buried in the rock. This example shows the importance of the family model in the cult. We can presume that in the first decades of the 4th century the Isiac cult was redeemed from the public. The connection with the Imperial cult had finally been broken and the Isiac cult became part of the opposition to the every day official Christianity. The followers of Isis from Salona created their closed, maybe even secret circles.

In the Archaeological Museum in Split another group of finds that belong to Salona is not fully complete. It is, however, generally known as a group of findings from the territory of central, maybe even southern Dalmatia. The museum in Split took possession of different private collections, the material of which was collected not only in the territory of Salona or on the islands of central Dalmatia, but also in the territory of Narona. We refer to four bronze statuettes of Osiris of the same iconographical scheme as that mentioned earlier, Osiris' statuettes,[70] fourteen *shabties*,[71] an amulet-pendant representing Isis *kourotropos* resp. *Isis lactans*,[72] an amulet-pendant representing the jackal Anubis,[73] an amulet-pendant representing the magic unharmed *wedjat* eye (Fig. 9/5),[74]

70 SELEM 1959, 94, no. 12-16; BUDISCHOVSKY 1977, 203, no. 22; SELEM 1997, 82, no. 2.34-2.37.

71 SELEM 1963-1965, 113-117; BUDISCHOVSKY 1977, 204-208, no. 24-32; SELEM 1997, 83-90, no. 2.38-2.51. For function see note 36.

72 H. 3.1 cm, Archaeological museum Split: SELEM 1959, 100, no. 20; BUDISCHOVSKY 1977, 209, no. 38; SELEM 1997, 90, no. 2.52.

73 H. 1.8 cm, Archaeological museum Split: BUDISCHOVSKY 1977, 210, no. 40; SELEM 1997, 91, no. 2.54.

74 H. 1.2 cm, wid. 1.3 cm, Archaeological museum Split: BUDISCHOVSKY 1977, 210, no. 42; SELEM 1997, 91-92, no. 2.55.

75 Both of height 6.5 cm, Archaeological museum Split, inv. no. H 358 and

two bronze statuettes of Hermes-Thoth,[75] and a statuette of goddess Nefertoum.[76]

The *shabti*, which in the ancient Egyptian language means "the answerer" answers, instead of the deceased, to perform any hard agricultural work. In the later periods their magical and metaphysical aspect increased. Their symbolism of death and regeneration connects them to the Isiac and Osiriac symbology, and they became instruments of the great act of the mystical renewal of the living strength of men. This idea was symbolised in this World by Isis and in the other World by her divine husband Osiris.

Diocletian's Palace in Split is above all a first-rate monument of classical architecture, today the nucleus of the city of Split.[77] This building was the last residence of *Caius Aurelius Valerius Diocletianus*, one of the most remarkable rulers of the Roman Empire. The precise year when the foundations of the Split palace were laid has not been historically established. The hypothesis that it was around 295 AD is based on the conclusion that Diocletian went forward with the undertaking some time after the tetrarchical system of rule had been adopted and after he himself had decided to retire. Ten years later the palace was to be ready to receive its master who, on May 1st 305 AD, laid aside the purple on a hill not far from Nicomedia and then set off towards his Dalmatian residence, not yet completely finished. In fact, a number of parts remained unfinished after the Emperor's death and remain in that condition even today.

The rectangular shaped Diocletian's palace occupies an area of about 28,900 square metres. The plan of the palace com-

1885 (both in Knin): ŽANIĆ-PROTIĆ 1988, 26-27, no. 16 and 18; SELEM 1997, 97, no. 2.64 i 2.63.

76 H. 5.5 cm, Archaeological museum Split. The divinity stands on a small lying lion dressed in the Egyptian apron, and on the head he has a lotus flower with two plumes: SELEM 1959, 104-105, no. 19.

77 BULIĆ-KARAMAN 1967; MARASOVIĆ-MARASOVIĆ 1970 ; etc.

bines the qualities of a villa and a fortified military camp (*castrum*). Its southern facade with the arched gallery on its upper floor rose directly from the sea. Three side-facades were fortified with sixteen towers, four of them placed in the angles. The monumental gate in the middle of each of these walls led to an enclosed courtyard. The dual nature of the architectural scheme, deriving from both the villa and the castrum type, is evident in the arrangement of the interior. The transversal street (*decumanus*) leads from the eastern to the western gates and divides the Palace into two parts. The main longitudinal street (*cardo*) links the northern gate with the southern section of the Palace. The northern part is thought to have been a large residential complex for servants and soldiers. The more luxurious buildings were built in the southern section of the palace. These included the monumental courtyard (*peristyle*) forming the northern access to the Emperor's apartments, a Mausoleum on the eastern side of the Peristyle and a complex of three temples on the western side, as well as two baths – an eastern and a western one. The Emperor's apartments formed the block along the seafront.

From the complex of Egyptian monuments a particular group belongs to the Diocletian's palace in Split. The Palace is permeated with eastern symbolism and as components of this symbolism appear the Egyptian sphinxes and the columns made of red granite from Assuan.[78] On the *peristyle* of Diocletian's palace the large sphinx is made of black granite with the body of the animal on a rectangular pedestal (Pl. 13/1).[79]

Particularities of this sphinx are that instead of the front lion paws it has hands which hold a vase (it is one of the biggest examples of the so-called *porte-vase* type sphinx) and also the frieze around the pedestal. In spite of damage, it has

78 Some of the columns are standing even today on the western colonnade of the Peristyle, the other are dislocated.

79 Lenght 246 cm, width 65 cm, height 100 cm: ADAM 1764, pl. LV-LVI; BULIĆ 1908, 98-99; JÉQUIER 1910, 175-179; BULIĆ-KARAMAN 1967, 81-82; SELEM 1970, 639-653; SELEM 1997, 98-100, no. 3.1.

been established that the frieze on the pedestal represents the usual geographical roll, where the bounded prisoners with cartouches sign up the subdued nations, places and countries. The production time of this sphinx is most likely the XVIII Dynasty, although there remain certain doubts in relation to its belonging to a certain ruler (Tuthmosis III, or possibly Tuthmosis I). The ruler's cartouche had been written on the vase, but it was destroyed in Egypt. The sphinx was usurped by Ramesses II, ruler of the XIXth Dynasty, whose cartouche remained visible.

Another great sphinx is made of white limestone resembling marble. Despite the damage it is obviously a quality work of Egyptian art (Pl. 13/2).[80] The head and the hind part of the pedestal are missing. The inscriptions are put on the chest of the sphinx, in a rectangular frame, and on the front, right and left side of the pedestal. The sphinx was made around the year 1400 BC, the ruler's name was Amenophis III. The sphinx made of shiny black granite is situated today in front of the Temple of Jupiter. Its head is missing and it also has damage to the front foot and on the pedestal Pl. 13/3).[81] It dates from the XXVIth Saite Dynasty. In addition, a fragment of a small sphinx made from white stone, a very damaged sphinx made of black granite, a small sphinx of grey granite, a fragment of a sphinx made from reddish granite, a fragment of a sphinx made of greyish-black granite, a small fragment of a sphinx also of greyish-black granite, an Egyptian pharaoh-head made from reddish granite (Pl. 14/2), a head of a sphinx made of reddish granite and a fragment of a sphinx of greyish granite have all been discovered here.[82] Thus at least eight sphinxes and another four Egyptian monuments have been attested, which shows Diocletian's inclina-

80 Lenght 151 cm, width 65 cm, height 100 cm, today in the Archaeological museum in Split: ADAM 1764, pl. LX; CONZE 1877; CONZE 1878; BULIĆ-KARAMAN 1967, 82-83; SELEM 1970, 639-653; SELEM 1997, 100-101, no. 3.2.

81 Lenght 180 cm, width 57 cm, height 74 cm: KARAMAN 1937, 615; SELEM 1970, 639-653; BUDISCHOVSKY 1977, 217, no. 57; SELEM 1997, 102, no. 3.4.

82 SELEM 1970, 639-653; SELEM 1997, 101-106, no. 3.3, 3.5-3.12.

tion to decorate the central spaces of his Palace with material of Egyptian origin. It can be connected to the ideas of the divine nature of the Imperial person, as the Egyptian sphinx is a symbol of sovereign strength. The Pharaoh with his divine personality connects this world with the world of gods and the sphinx watches the roads that lead to the Underworld, also showing the power of the ruler. The ruler's honour at the time of Diocletian received expressive eastern characteristics, among them concerning his divine nature, although the first Roman Emperors had already had the epithet *divus* - divine. But this was mostly the way of declaring the magnificence of their person, the strength of their virtues, which can not be compared to the fact of being a god's son or a god's personality of the eastern rulers. Diocletian was formally Jupiter's son, *Iovius*, as it stands in his official nomenclature. The sphinxes are regularly discovered in the vicinity of the Peristyle and Diocletian's Mausoleum.

From the time of the earliest contacts between Dalmatia and the Greek world the river Narenta (Neretva, *Naro flumen*) was a major trade route between the Adriatic and the interior. Before the fourth century a Greek trading post (*emporion*) was established near the mouth of the river. Its importance increased with the growth of the settlement ***Narona***. The ruins of the ancient town are located in the village of Vid some 4 kilometres to the west of Metković. The town was situated on the hillsides, and the Neretva River forms a natural protective wall. It was built according to a Graeco-Illyrian urban concept, spreading in the form of a fan towards the east. Its eastern flank remained unfortified. *Narona* frequently served as a Roman military base for the campaigns against the Dalmati. In the late Republican period it was the seat of the *conventus iuridicus*, the administrative and judicial centre of the wider region. During the rule of Emperor Augustus, Narona was granted the status of a colony, and this event marked the beginning of the period of its greatest prosperity.

83 Alabaster, height 28 cm, height with the base 32.5 cm, Archeological collection Vid: CAMBI 1963-65, 98-100; BUDISCHOVSKY 1977, 187-188, no. XIV, 2; SELEM 1997, 112-113, no. 5.2.

The cult of Isis in Narona has been witnessed due to the finding of the bust of the goddess, made from alabaster, on the site Bostan-Bare (Pl. 14/1).[83] The bust lacks the face, while the rest is quite well preserved. The face had to be inserted in a specially made camp (height about 13 cm, thickness about 2-3 cm), forming a veil which would frame the face of the goddess. The camp was hollowed with a chisel without further polishing. The soft alabaster would not be suitable for frequent changes, so it should eliminate the possibility of the bust holding the face of goddess' priestesses, because with every new priestess the face would have had to have been changed. The bust lacks some of the usual attributes (the lotus flower, the crown), but one of them was certainly attached above the front, near the edge of the veil where the remains of an iron wedge used for fixing it are visible. The edge of the veil and the scarf are nicely decorated.

It is without doubt the representation of the goddess Isis, dressed in her usual clothes – a richly draped *peplos* with a scarf thrown over and decorated with fringes. It is tied in a knot at the front. The missing face was made of alabaster or some other sort of stone, most probably black granite, which is frequently the case. The black face of the goddess or the black dress symbolically declares the goddess's grief for Osiris. By the treatment of the drapery the bust can be dated to the end of the 2nd or the beginning of the 3rd century AD. The rear was treated summarily which indicates that the bust was put in a niche. This bust is a certain testimony of the veneration of this Egyptian goddess in Narona and it proves the presence of an Oriental population. A tombstone from Narona decorated with two *sistrums* in relief is further proof.[84] Furthermore, from Narona comes a small circular bronze plate representing a male head with luxurious hair, beardless, but with horns and ram's ears. Therefore it could

84 CIL 3, 1864; BUDISCHOVSKY 1977, 187, no. XIV, 1; SELEM 1997, 112, no. 5.1.

85 BRUNŠMID 1914, 248, no. 165; BUDISCHOVSKY 1977, 188; SELEM 1997, 113, no. 5.3.

be the representation of Jupiter-Amon.[85]

The fourth colony on the Dalmatian coastline bore the Greek name ***Epidaurum***. The small headland south of Dubrovnik may have been a trading post frequented by merchants, but there is no evidence that it was ever an established Greek settlement. The place is first recorded as a stronghold (*praesidium*) of Caesar during the Civil War, attacked by the Pompeian fleet in 47 BC but saved by the intervention of Publius Vatinius.[86] The foundation of the colony (enrolled in the tribe *Tromentina*) was probably part of Caesar's program to reward his allies in the Adriatic. *Epidaurum* may once have been a great coastal town, but in late antiquity parts of it were submerged beneath the sea.[87] There is no evidence of the original size of the town, but it is unlikely that it compared with any of the larger colonies to the north.

In the year 1890 a bronze statuette of the goddess Sekhmet was bought in Dubrovnik (Pl. 14/5).[88] The exact place of finding is not known, but presumably it comes from the surroundings of Dubrovnik, most probably from Cavtat, i. e. the Roman *Epidaurum.* The goddess is represented as a naked woman with the head of a lioness. On her head she is wearing the *klaft* with the sun-disc and the *ureus* in front of it. The elaboration is simple. Sekhmet, whose name means the Powerful, was a member of the Memphite triad. She was the mistress of fire, of vital forces and of the sky. She doesn't appear in Europe with the divinities of the Isiac circle. Most probably this example, a fine work of Egyptian handicraft of the Late period, was also imported from Egypt, maybe as the protective amulet of a travelling surgeon.

Pharos (Stari Grad, the island of Hvar). The ancient history of Hvar was written in Greek and Latin, mostly by geographers and

86 Caesar, *Bellum Alexandrinum*, 44.

87 Along the southern part of the Dalmatian coast the land appears to have subsided more than 2 metres in places.

88 Bronze, height 11.9 cm, Archaeological museum Split: SELEM 1959, 106-197, num. 17; BUDISCHOVSKY 1977, 208, br. 33; SELEM 1997, 115-116, no. 5.5.

historians. A great deal of information dates from the 4th century when the Greeks formed a settlement on the island. Ionic Greeks from the island of Phar settled the island with the help of Dionysius the Elder, the tyrant of Syracuse, between 395-393 BC. The intervening period seems to have been taken up with fighting against the original inhabitants, who were already settled in the place where the Pharos newcomers wished to build their settlement. Pharos is shown as an important settlement of Illyrian seamanship during the First and the Second Illyrian War. From this we can suppose that the Greek settlement on Hvar had a mixed Illyrian-Greek character. The settlement was destroyed by the Romans during the Second Illyrian War in 219/8 BC. Its fate is learned from inscriptions, coins and various other sources, from remains of the settlement itself, such as the ancient division of the Stari Grad field system. We discovered that *Pharos* continued to exist, or to be more precise that it was reborn as a Greek settlement, then again it came under the local ruler and was finally restructured following the Roman pattern (*municipium)* at the beginning of the 1st century AD. Pharos possesses a rich and particular inventory of small sculptures of Egyptian origin.[89] They are statuettes of divinities or amulets, all connected to the Isiac circle.

To this collection belongs a bronze statuette of Osiris, iconographically equal to the Osiris statuettes from Salona men-

89 BUŠKARIOL 1988, 37-39.

90 Height. 6 cm, Archaeological museum Split: SELEM 1959, 94-103, no. 14; SELEM 1994, 453, no. 1; SELEM 1997, 107, no. 4.1

91 Height 12.2 cm, the figure is standing with the left leg largely distretched. On the head he has the crown with the lotus flower and two long plumes. The lower part is lost but probably he stood on a lion: SELEM 1994, 455-456, no. 8; SELEM 1997, 107-108, no. 4.2.

92 Height 5.5 cm, Archaeological museum Split. The divinity stands on a lying lion, on the head he has the crown with the lotus flower and two plumes, on the body only an Egyptian apron: SELEM 1959, 104-105, no. 19; SELEM 1994, 456, num. 7; SELEM 1997, 108, br. 4.3.

93 Width 3.8 cm, Archeological museum Split. The crown *psent* on the head of the bird determines this falcon as the divine Horus: SELEM 1997, 108-109, no. 4.4.

tioned earlier (Pl. 14/4),[90] Furthermore a bronze statuette of the god Nefertum (Pl. 14/3),[91] a green faience pendant representing Nefertum,[92] an amulet-pendant in the form of a falcon,[93] an Osiriac statuette similar to a *shabti*,[94] a *shabti* figure,[95] two amulets in the form of the multiple *wedjat* eyes and the simple *wedjat* eye,[96] a gem made of reddish carnelian with the busts of Isis and Serapis (Pl. 10/643)[97] belong to the collection. Nefertum, as the lotus god, in the Late Period of the Egyptian culture takes the role of the god's son in the Osiriac triad, sharing this position with Horus / Harpokrates. He wasn't particularly widely spread throughout the European provinces of the Empire. The amulet-pendant in the form of a falcon is a nice example of Egyptian handicraft of the late period. The appearance of the royal Horus in the form of an amulet of prophylactic use is quite widely spread in the creed of the Hellenistic-Roman period.

The monuments of Egyptian cults in Dalmatia, according to the known findings, are almost exclusively spread along the coastline and almost unknown in the hinterland. Firstly, we will mention here the finding of a bronze statuette of Isis-Fortuna from the territory of ***Lika*** (Pl. 15/1).[98] The goddess is standing, dressed in a chiton and chimation fastened around the waist. On her head she has a complex Isiac crown: the moon disc is surrounded by two cow horns, with two corn ears in the middle. On the front there is an *uraeus*. The right arm leans against a tiller, while the left hand is holding the corn of abundance. Isis-Fortuna is the protector of navigation, she is the goddess which brings

94 BUDISCHOVSKY 1977, 185; SELEM 1994, 456, no. 3; SELEM 1997, 109, no. 4.5.95 H. 4.8 cm, Archaeological museum Split: BUDISCHOVSKY 1977, 186; SELEM 1994, 456, num. 2; SELEM 1997, 109-110, no. 4.6.

96 I. dim. 4 x 4.1 cm, II. dim. 2 x 3 cm, Archeological musem Split: BUDISCHOVSKY 1977, 186-187; SELEM 1994, 456, nos. 5 and 6; SELEM 1997, 110-111, nos. 4.7 and 4.8.

97 PHAROS 1995/96, no. 3.

98 H. 7.3 cm, Archaeological museum in Zagreb: BUDISCHOVSKY 1977, 178; SELEM 1997, 48-49, no. 1.3.

luck and abundance. The tiller, i. e. the rudder in her hand reminds us of the most important Isiac public festivities in the Hellenistic and Roman period, called the *thalassai*, i. e. the *Navigium Isidis*, which began on the 5th of March (later the 3rd of January).[99] Owing to Apuleius,[100] we have the description of these Isiac festivities with which the season of navigation began under the protection of the goddess Isis.

In ***Oklaj*** near Promina (***Promona***) a bronze vessel in the shape of the god Bes was discovered (Pl. 15/2).[101] The grotesque divinity is represented in his typical form – a wide head, luxurious hair, moustaches and a beard, a grinning face. The lion skin covers the upper front part of the body, while the lower part of the corpulent belly and the genitals are uncovered. The feet end with animal paws. Above the head there is the aperture upon which the cover is placed. At the height of the shoulders there are two apertures through which the rings of the handles are pulled. The handles end in the form of swamp birds. It is a fine work of handicraft of the Hellenistic-Roman period and it can be dated to the 1st century AD. This type of vessel has two points of origin, the Egyptian vessels with the figure of Bes and the Hellenic vessels with the figure of Silen. The animal paws are Silen's characteristic attributes.[102] Bes appears as the companion of female divinities of the Isiac and Hathoric circle.

From ***Prolođac*** near Imotski comes another bronze statuette of Isis-Fortuna (Pl. 15/3).[103] The goddess is represented

99 ALFÖLDY 1937; TIBILETTI 1997.

100 Apuleius, *Metamorphosis*, XI, 8-17.

101 Bronze, height 7 cm, Archeological Museum Split, inv. H.4427: BULIC 1909, 47; BISSING 1925, 123ss; BULIC 1926-27, 105; BUDISCHOVSKY 1977, 183-184; SELEM 1997, 92-93, no. 256.

102 A vessel from the necropolis Mrtvilo on the island of Vis (*Issa*) which represents Bes-Silen has been recently published with just a photography. It is a work of the Alexandrian workshops of the 1st century A.D.: BRUSIĆ 2000, 530.

103 Height 7.9 cm, width. 2.7 cm, Narodno sveučilište Imotski: TONKOVIĆ 1991, 53-54; SELEM 1997, 93, no. 2.57.

standing, dressed in a long dress with the *palla*. In her left hand she holds the corn of abundance, full of different fruits with a crescent on the top. The right arm certainly leant against a tiller, which is missing today. The *kalathos* forms the only cap of the goddess. The hairstyle dates this statuette to the beginning of the 2nd century AD.

In the Collection of the Franciscan monastery in ***Sinj*** two terracotta *shabti* figures from an unknown site (possibly Čitluk in the vicinity of Sinj – *Colonia Claudia Aequum*) are kept.[104]

Outside the territory of the ancient Liburnia, but still in Dalmatia, an early testimony of the appearance of the Isiac cult amongst the natives exposed to Romanization is the cippus from ***Ljubuški Humac*** (***Bigeste***).[105] In the right upper corner the *sistrum* is visible. The cippus has been dated to the Flavian period and brings out a native family in the process of Romanization as the carrier of her cult.[106] On this monument the connection of a native family with the Isiac cult is documented with the carved out sistrum.

Special types of monuments are the votive and dedicatory inscriptions where the name of one or more gods appears. They are the only firm and direct proof of the religious belief of the person who erected them.[107] But it is necessary to separate the inscriptions with theophoric names and the inscriptions with figural representations connected to the cult of Isis. Eight inscriptions are known from the Dalmatian coast with theophoric names. They are: the inscription with the theophoric name of

104 BUDISCHOVSKY 1977, 184; SELEM 1979, 80, no. 2 and 3; SELEM 1997, 94-96, no. 2.58-2.59.

105 Great rectangular cippus of white marble, dim. 111 x 143 x 80 cm, Lands Museum in Sarajevo: CAMBI 1963-65, 105; RENDIĆ-MIOČEVIĆ 1989, 632; SELEM 1997, 114-115, no. 5.4.

106 SELEM 1972, 88-89.

107 Most of all of those are votive altars (ara), then sepulchral monuments, pedestals of statues, plates of votive character. The inscriptions can be accompanied with figurative relief representations.

Isidora from Zadar or Nin,[108] a fragment of a tombstone with the name ***Apius*** from Obrovac (*Clambetae* ?),[109] the tombstone from Trogir or Klis near Salon with the name ***Sarapia***,[110] the tombstone from Salona with the name ***Serafio***,[111] the tombstone from Split (?) with the name ***Isicia***,[112] a marble sepulchral plate from Salona (?) with the name ***Isidorus***,[113] a stele from Dugopolje (road Split-Sinj) with the name ***Isiada***,[114] a stele from Pučišća on the island of Brač (*Brattia*) with the name ***Serapion***.[115] The rule, corroborated in Italy, that the Egyptian theophoric names mostly appear as *cognomina*, is confirmed in Dalmatia. The big question is, however, in what proportion theophoric names state the pertinence to a certain cultural area. Though not certain, it is therefore reasonable to presume that these persons were of Egyptian origin and they remained in a certain way in touch also with the Egyptian religion. The types of objects on which figures of Isis and other figures of the Isiac circle appeared were at first the monumental statue and bust which are testimonies to the public and official cult. On the other side there are numerous smaller objects, statuettes, amulets, lamps, gems that prove the personal or private level of the cult.

The Dalmatian coast was open to the great maritime paths from the 4th century BC, when the Greek colonization began. If we take into consideration a comparatively large number of small objects, statuettes and amulets of Isiac content and Egyptian ori-

108 CIL 3, 10004; BUDISCHOVSKY 1977, 176; SELEM 1997, 50, 58, no. 1.5 and 1.16.

109 COLNAGO-KEIL 1905, 2, 46; BUDISCHOVSKY 1977, 182; SELEM 1997, 59, no. 1.17.

110 CIL 3, 2401; BUDISCHOVSKY 1977, 184; SELEM 1997, 62, no. 2.1.

111 BULIĆ 1910, 30; BUDISCHOVSKY 1977, 193, no. 4; SELEM 1997, 80-81, no. 2.31.

112 SELEM 1997, 81, no. 2.32.

113 CIL 3, 8935; BUDISCHOVSKY 1977, 193, no. 4; SELEM 1997, 81, no. 2.33.

114 CAMBI 1993, 164-165, no. 7, fig. 12; SELEM 1997, 95-96, no. 2.60.

115 CIL 3, 3103; BUDISCHOVSKY 1977, 185; SELEM 1997, 112, no. 4.10.

gin discovered in Pharos, the question arises logically whether the Isiac cult came with the Romans on this coastline, or whether some of its signs had already arrived before that. We are sure about are the Italic connections through Aquileia and that the great harbours on the Eastern Adriatic coast were effective in the 1st and 2nd century AD. In the 1st century Egyptian cults were present in all of the classes of the social structures, while at the same time in Rome, without looking at some singular and isolated examples, the Egyptian cult slowly rose from the lower classes of society. At the end of the 2nd and the beginning of the 3rd century, during the Severan dynasty, a new regeneration of the cult can be considered. It is the period of Serapis' rise, the time of great assimilation, sincretism and contamination among the eastern creeds. At the end of the 3rd and the beginning of the 4th century Christianity slowly took the positions once held by the eastern and Egyptian cults. The Isiac cult began to close up in smaller, aristocratic groups, as we saw the example in Salona, which became the forts of resistance against Christianity. The last ascension was at the time when the Emperor Diocletian decorated his Palace with Egyptian sphinxes and other monuments.

The appearance of Egyptian cults on the territory of the Roman province of Dalmatia is articulated by the very wide radius of their spread through the entire territory of the Roman Empire and even beyond its borders. The diffusion of the Egyptian gods in the Mediterranean certainly did not begin with the Ptolemaic period. Syrian and Palestine territories were under strong Egyptian cultural and religious influence from older times, Greece also received the same influence in the pre-classical phase. Moreover, it is sure that the political strength and cultural fascination of Hellenistic Egypt gave the final impulse to this diffusion. Isiac traces are numerous throughout this territory, from popular amulets and statuettes to monumental marble statues and Pharaonic sphinxes. These traces are equally connected with the slaves and libertines as with the citizens, the *eques* and the Imperial person, decorating his Palace with Egyptian objects. Isis' followers were certainly in minority in the *municipia* and colonies, but their presence radiated with a special spiritual and cultural charge, acting also outside the severe cultural frames.

The eldest monuments from Dalmatia originate from the second half of the 1st century AD. They are the statue of Isis from Nin and the cippus from Bigeste with the representation of the *sistrum*, most certainly from the Flavian period. But the context of these monuments raises the possibility that at the time of their appearance, the Isiac cult was not an entirely new religious and cultural factor but already had a certain tradition. The statue of Isis from Nin is a testimony of the presence of an already developed cult of the goddess on the territory of Liburnia, not far away from the time of its legalization in Rome. The later inscription from Zadar proves that the Isiac cult on the Liburnian territory followed the general development. Isis on the altar from Zadar doesn't appear alone, but with her companion Serapis, although she is still more important and her name appears first in the dedication.

The receptiveness of the Liburnians toward the mighty goddess found its base firstly in the series of female divinities by the Liburnians, which obviously had many similarities with the universal Hellenistic Isis.[116]

116 According to inscriptions from the Roman period, the Liburnians worshipped predominantly female deities in the old Mediterranean matriarchal tradition. Almost every territorial community (*civitas* later *municipium*) had its local deity. *Anzotika* is known from Nin, called and represented in the form of the Roman Venus (with the little Priapus). She preserved the local name and most certainly represented the mother-goddess in her aspect of fertility. E.g. The sculpture of Venus Anzotika from Nin, marble from Carrara, height 100.5 cm, Archaeological museum Split: DEGRASSI 1937; GALLI 1939; ABRAMIĆ 1939; ABRAMIĆ 1940, 174-176; CAMBI 1980a; GIUNIO 1997, 239-242. Two inscriptions with her name were found in Nin, in one of them this goddess corresponds to Venus and the inscription was dedicated to *Venus Anzotica*. This is an instance of the so-called Roman interpretation of native cults (*interpretatio romana*). It is certain that in the minds of the natives Anzotica was not the goddess of beauty like the classical Aphrodite or Venus but rather the female principle of fecundity and fertility, the earthly and the cosmic symbol of reproduction of all living beings, the source of living power. The lack of figurines of Liburnian deities is probably due to the lack

The *ara* from Bribir and the stele from Asseria are the proof that the Liburnians themselves joined the Isiac religion or at least were touched by its culturological radiance. *Raecia Marcella* originates from a respectable family from the Liburnian territory and also the family of *Vadica Titua.*[117] The acroteria with the figures of Jupiter-Amon on Vadica Titua's stele are most certainly proof of the Egyptian cultural influence, but, as already mentioned, Jupiter-Amon appears only on the edges of the Isiac circle, only his Egyptian origin connects him to it. His symbolism and significance is autonomous, thus Vadica Titua, and the others mentioned in the inscription on her funeral stele, were most probably just touched by the Egyptian cultural influence. Likewise, Jupiter-Amon could appear here just in connection with some sepulchral symbolism, having nothing to do with his cult.

Jupiter-Amon, with the status of a supreme or state god appears mostly as a decorative motif in combination with the figure of Meduse. Iconographically he is represented as a bearded man with ram's horns. He is a product of Hellenism and so his iconography belongs to its artistic area. His figure can be found on a lamp from Nin and on an applique or phalera or cover from Salona (mentioned earlier).[118] As a particular group among his representations, I would separate the appearance of his figure in the architectural decoration of

of tradition of personification of deities, a custom which was first brought to the Liburnians by the Romans.

117 Monumental funeral stele made of white limestone from Podgrađe near Benkovac (Roman *municipium* of Liburnian tradition *Asseria*), dim. 253 x 67 x 18 cm, Archaeological museum Zadar, inv. no. 28. Consists of a gable, niches with portraits of two female persons, the profilated *titulus* and the representation of the funeral doors in relief. On the gable in the middle is represented the *gorgoneion*, on the sides figuredthe acroteria like two heads of a horny divinity in profile – Jupiter Amon: BUDISCHOVSKY 1977, 183; RENDIĆ-MIOČEVIĆ 1989, 578-580; SELEM 1997, 61-62, no. 1.19.

118 Belonging to an unknown site a head of Jupiter Amon kept as a console or keystone of the towngate (height 23, width 17, thickness 17). In the Archaeological museum in Split: CAMBI 1980, 43-45.

the Forums in Pola and Zadar, as part of a general phenomenon in the decoration of a Forum of the north Adriatic – Aquileia, Zuglio, Trieste, Pola and Zadar.[119] Four monolithic limestone blocks from the upper part of an altar enclosure come from Pola. On the front side, in a rectangular *aedicula*, the head of Jupiter-Amon is chiselled in high relief (Pl. 15/4 and 16/1-4).[120] Two pilasters with the heads of Jupiter-Amon and a pilaster with the head of the Medusa come from Zadar. They belong to the upper floor of the loggia, the so-called *basilicae Severianae* situated on the Forum (Pl. 17/1-3).[121]

The common appearance of Meduse and Jupiter-Amon on the Forums of the Roman cities of the north Adriatic was explained by M. C. Budischovsky as the glorification of the honour and strength of the Roman army on land and sea. We also find this iconographic tandem in the Roman cities of Spain, Gaul and Germany.[122] The phenomenon could be in some way connected with a religious purpose. Jupiter-Amon could be directly connected to the strength of the Roman army, i. e. the Egyptian dynastic god has been merged into the Roman Stately - Imperial religion and cult with the supreme Roman divinity. Alexander the Great was already identified with Amon and his figure frequently appears with ram's horns on the head. He became the ideal ruler with whom all the early Emperors tried to be identified, thus the iconography of Jupiter-Amon appeared for the first time officially

119 BUDISCHOVSKY 1973, 201-220.

120 Three blocks are from the 1st century, of appr. height around 120 cm, width around 80 cm, thickeness around 85 cm, Archaeological museum of Istria, Pula. The fourth block is dated in the 2nd or 3rd century, of similar dimension: JURKIC-GIRARDI 1970, 27-43; BUDISCHOVSKY 1973, 204-206; BUDISCHOVSKY 1977, 171-173; JURKIĆ-GIRARDI-DŽIN-URANIĆ 2001, 8-12.

121 Limestone, height 132 cm, width 100 cm, thickness 101 cm: SUIĆ 1965, 201-203; BUDISCHOVSKY 1977, 179-180; SELEM 1997, 51-53, no. 1.7 and 1.8; GIUNIO 1999, 60-61.

122 BELLIDO 1949, 415; VERZAR 1978, 33-46.

123 GANZERT-KOCKEL 1988, 149-199; SIMÓN 1990, 143-162.

in the decoration of Augustus's Forum in Rome.[123]

Oriental cults in Roman dominated territories appeared at the historic time conditioned and determined by important social factors as well as by socio-political and philosophic-religious conditions in the Roman state. The need for new deities in the popular religions was based on ancient autochthonous beliefs of the conquered population. They appeared in the form of ideas that had gradually come into shape and followed almost continuously the strengthening of the Roman empire during the shift from the Republic to the Empire.

The presence of the above mentioned Egyptian and Roman-Egyptian deities confirms a permanent existence of commercial, artistic and religious ties between the south-eastern Mediterranean and the Croatian coastal area that made it possible for the newly-arriving populations to coexist and cultivate their traditions in the period from the 1st to the 3rd century AD.[124]

Kornelija A. Giunio
Archaeological Museum of
Zadar

124 At the end I would like to express my profound love and gratitude to my mother Nikolina for her love and patience. Furthermore, I would like to thank my colleagues Josipa Gulin for the lecture of this English text, Ivana Mazić for the drawing of the Map, and for their great patience. Many thanks to the Hungarian National Museum and The Ancient Egyptian Committee of the Hungarian-Egyptian Friendship Society for accepting my contribution.

LES TERRES CUITES D'HARPOCRATE-AU-FAUCON-SUR-LE-PILIER

Harpocrate, Horus l'enfant, était un parmi les dieux importants du culte alexandrin dans l'empire romain mais peu fréquemment mentionné dans les sources écrites. Il était vénéré principalement en tant que fils d'Isis et/ou de Sérapis dans de nombreux pays du monde hellénistique. Sa place de second plan dans le culte s'est également reflétée par le nombre moins élevé de ses représentations, seul ou en compagnie des dieux mentionnés.

En Égypte, son pays d'origine, l'image du dieu est un peu plus nuancée. Horus, adulte ou enfant, a continuellement joué un rôle central dans les cultes traditionnels des temples gréco-romain. En tant que dieu enfant, il incarnait le renouvellement de la vie. Les papyri magiques suivent la tradition ancienne et lui attribuent un pouvoir de guérison, à l'instar d'Horus sur les crocodiles, domptant les animaux maléfiques, représenté sur les stèles. Dans les couches sociales élevées de la population de l'Égypte hellénisée, la vénération d'Horus l'enfant est aussi important,[1] mais son rôle est encore le plus souvent liée à Sérapis et à Isis[2], comme dans les autres régions hellénisées. Par contre, ses représentations (seules et liées aux autres dieux) en terre cuite[3] ainsi qu'en bronze dépassent en nombre même celles de la déesse Isis, montrant ainsi qu'il était le plus populaire des dieux(-enfants) durant la période ptolémaïque et la période romaine dans

1 NACHTERGAEL 1995.

2 NACHTERGAEL 1995.

3 Voir les catalogues de terres cuites et DUNAND 1984, 21; TRAN TAM TINH 1988; BALLET 1988 – autres travaux importants que je ne pouvais pas consultés: POULIN 1987; COONEY 1972.

un milieu hellénisé socialement moins élevé. Fabriquées pendant plus d'un demi-millénaire, les terres cuites illustrent des épisodes de la vie du dieu et ses caractères, de même qu'elles le rapprochent des autres dieux-enfants d'origine égyptienne ou grecque.

La majorité des terres cuites représentant Harpocrate ou le cercle de son culte dont la provenance est connue ou probable[4] est originaire de la Basse Égypte. Les ateliers les plus importants et les utilisateurs se trouvaient dans le Delta[5], à Alexandrie. Cependant, les terres cuites d'Harpocrate et harpocratiques étaient aussi très populaires auprès des habitants d'Aboukir[6], de Naucratis[7], de Kom Firin[8], de Terenouthis[9], de Saïs[10], de Mahalla el-Kubra[11], d'Achmoun[12], de Tanis[13], de Thmouïs[14], de Tell Roba/Mendes[15],

4 Les publications des découvertes archéologiques récentes sont attendues: voir NACHTERGAEL 1995a.

5 Dans quelques cas seulement la provenance donnée est "Basse Égypte". Par exemple PERDRIZET 1921, n° 84, 91, 98, 107, 112, 119, 127, Le Caire JE 38164, JE 38260 = CG 32998.

6 BALLET 1988, doc. 182= BRECCIA 1930, 69, pl. 35/8, doc. 183, doc. 203= BRECCIA 1930, 69, pl. 35/7, doc. 339, doc. 340= BRECCIA 1930, 69, pl. 35/2, doc. 341, doc. 389= BRECCIA 1930 , 70-71, pl. 36/1, BALLET 1988, doc. 415= BRECCIA 1930, 69, pl. 35/1, doc. 427-428, doc. 447= BRECCIA 1930, 69, pl. 35/4, doc. 501= BRECCIA 1930, 70, pl. 35/9,

7 GUTCH 1898-1899, 67-97, pl. XIII; DUNAND 1990, n° 252, Le Caire CG 43445 = 14-9-18-4, CG 32822, cf. PETRIE 1886, 45.

8 PERDRIZET 1921, n° 90, pl. 22, ŕ droite en haut, n° 95, pl. 35, ŕ gauche en haut, n° 124, pl. 34.

9 Le Caire JE 91815, 91816.

10 Le Caire CG 26643 = JE 32797.

11 DUNAND 1979, n° 262 = Le Caire 67868.

12 PERDRIZET 1921, n° 93, pl. 23, ŕ gauche, en haut, n° 102, pl. 27, ŕ gauche, en bas.

13 DUNAND 1990, n° 107, 218, 258, 276, 312, 313, 314.

14 DUNAND 1979, n° 153 = Le Caire 56832, n° 278 = Le Caire 36431.

15 BRECCIA 1934, n° 75, p. 24, pl. 26/119 = MEULENAERE - MACKAY 1976, n° 206, p. 216; BRECCIA 1934, n° 75, p. 24, pl. 15/61 = MEULENAERE - MACKAY 1976, n° 207, p. 216.

de Kantara[16], de Qasr Gheit[17], de Tukh el-Qaramus[18], de Boubastis[19], d'Athribis/Benha[20], d'Avaris/Tell el Daba'a[21]. Ces statuettes ont été retrouvées en très grand nombre dans la région du Fayoum, dans les villes de Dionysias[22], Karanis[23], Philadelphia[24], Theadelphia[25], Narmouthis[26], Sunufar[27], Hérakléopolis[28], Hawara[29], el-Wasta[30] et Arsinoë/Krokodilopolis[31]. Les terres cuites marquées de Medinet el-Fayoum[32] et de Kîman Farès[33] apparti-

16 WEBER 1914, n° 48, pl. 4, n° 73, pl. 6, n° 81, pl. 6, n° 125, pl. 12; FROEHNER 1867, 17; VANDIER 1971, 96, note 6.

17 CLÉDAT 1912, 164.

18 Zagazig, Musée Orabi, reg. n° 584, 587.

19 WEBER 1914, n° 40, fig. 58, n°. 53, pl. 4, n° 145, pl. 14, Zagazig, Musée Orabi, reg. n° 569, 575, Le Caire CG 43515 = JE 38811.

20 PERDRIZET 1921, n° 110, pl. 51, en bas, n° 85, pl. 26, au milieu; BAYER-NIEMEIER 1988, n° 167, pl. 34.1, Zagazig, Musée Orabi, reg. n° 579, MYSLIWIEC-RAGEB 1992, 415-416, fig. 14c, 15.

21 BIETAK 1968, 105, pl. XXIIc.

22 SCHWARZ-WILD 1950, 93, n° 14-58, pl. 22.

23 ALLEN 1985, n° 45-64, Le Caire JE 65602, JE 65569, CG 65602, 65597.

24 WEBER 1914, n° 98, pl. 9.

25 WEBER 1914, n°. 60, pl. 5, n° 70, pl. 6, n° 76, pl. 6, Le Caire, CG 41033, Musée d;Alexandrie 19689 (557A).

26 BRESCIANI 1968, 29, n° 13 (Isis lactans), 19 (monté sur un cygne) = pl. VII. 1, VIII.2, p. 53, n° 37; 38 = pl. LIX.3 = C.1, LIX.4; BRESCIANI 1976, 13. n° 26, 27, 35 (chevauchant), 36? (chevauchant), pl. V, VII.

27 DUNAND 1979, no. 167, 193, 198, 215-6, 232, 242, 260, 328, Le Caire JE 60586

28 PERDRIZET 1921, n° 116, pl. 26, à droite, en haut; SKUPINSKA-LOVSET 1978, UT 122, Oslo 39802e; PETRIE 1905, 1-2, et 26-27, n° 37-44, pl. XLVII-XLVIII.

29 ALLEN 1985, II. n° 52 parmi les comparanda.

30 BAYER-NIEMEIER 1988, n° 124, p. 101, pl. 23/6.

31 WEBER 1914, n° 112, pl. 10.

32 BAYER-NIEMEIER 1988, n° 30, 90, 144-47, 148-56, 162, 164; WEBER 1914, n° 50, pl. 4, n° 67, pl. 6, n° 102, pl. 9.

33 DUNAND 1979, n° 186, 224, 228-9, 274-5, 277, 283, 293, 326-7, Le Caire,

ennent à cette même ville. Ajoutons encore les nombreuses terres cuites dites simplement provenir du "Fayoum".

La région située entre le Delta et le Fayoum a également donné de nombreuses statuettes en terre cuite appartenant au culte d'Harpocrate, notamment les sites de Memphis[34], Mit Rahineh[35], Saqqarah[36], Tell el Nawa[37] et Gizeh[38]. En Haute Égypte, à l'exception d'Antinoé/Antinoupolis[39], les trouvailles sont moins nombreuses. Les inventaires mentionnent: Touna el-Gebel[40], Hermopolis[41],

JE 59241, JE 59242, JE 59244, JE 41609, JE 42031 = 25-11-22-31, JE 89089= 30-10-48, JE 42029 = CG 43243, JE 42046 = CG 43249, JE 42052 = CG 43259, JE 42019 = CG 43260, JE 42053 = CG 43262, JE 42047 -7 CG 43263, JE 42051 = CG 43264, JE 42050 = CG 43268, JE 42083 = CG 43278, JE 39331 = CG 43426, JE 39330 = CG 43527, JE 39330 = CG 43527, Musée de Kom Ouchim, reg. n° 38/1, 42/2, 45, 69, 94, 117, 129, 144, 246-47, 253, 272, 347-48, 373, 529, 540, 636/5-7, 638, 578, 687, 749a-b, 758.

34 PETRIE 1910, 45; ENGELBACH 1915, 34, pl. LXI/45; WEBER 1914, no. 150, pl. 14; PERDRIZET 1921, n° 81, pl. 23, à droite, en haut, n° 82, pl. 25, à droite, en haut, no. 97, pl. 23, à droite, en bas, no. 100, pl. 27, à gauche, en bas, no. 101, pl. 27, à droite, en haut, no. 103, pl. 35, à droite, en haut, n° 122, pl. 26, à droite, en bas, n° 123, pl. 34, au milieu, n° 125, pl. 34, à droite, en bas.

35 BAYER-NIEMEIER 1988, n° 47, pl. 9.1; DUNAND 1979, n° 189 = 25-11-22-1, no. 201 = 32868, Le Caire CG 43428, CG 43517 = 31-7-27-9, 25-1-22-4.

36 DUNAND 1979, n° 134 = 27003, n° 282 = 33300 = 43463, n° 190 = JE 6053 = CG 26980, Le Caire JE 51737 = CG 43418, JE 6029 = CG 26996, JE 6040 = CG 27097, JE 91941, JE 91312.

37 DUNAND 1979, n° 184 = p. 241.

38 WEBER 1914, n° 96, pl. 9.

39 DUNAND 1990, n° 115, 126, 129, 137-8, 146-7, 157, 165-6, 168, 174-180, 182, 185, 200, 211-12, 235, 253-55, 273, 274-75, 326; BESQUE 1992, 108, pl. 66e, p. 111, pl. 70b, p. 112, pl. 70d.

40 PERDRIZET 1921, n° 118, pl. 30, à droite, Musée de Mellawi, reg. n° 175, 414, 415, 417/3.

41 PERDRIZET 1921, n° 94, pl. 23 au milieu, n° 115, pl. 91, à droite en bas, Le Caire JE 65454.

42 BALLET 1988, doc. 339, 599-621. – les fouilles de Reinach à 1911, voir

Koptos[42], Tentyris[43], Médamoud[44], Medinet Habou[45], Louxor[46], Thèbes[47], Apollinopolis[48], Douch[49] et peut-être Assouan[50] (indication incertaine). Bien que plusieurs terres cuites aient dû être importées du Nord, des ateliers locaux fonctionnaient aussi dans le Sud et fabriquaient des pièces caractéristiques.

Les représentations les plus fréquentes du culte d'Harpocrate sont des terres cuites et des statuettes en bronze représentant le dieu dans des poses variées et avec différents attributs. Parmi les statuettes en bronze, les représentations d'Harpocrate sur le trône et debout tenant la corne d'abondance, sont les types les plus fréquents.

L'iconographie des terres cuites est riche aussi. Les types les plus populaires sont Harpocrate enfant au pot[51], en général assis, mais quelquefois aussi debout, et Harpocrate adolescent avec la corne d'abondance, le plus souvent debout et plus rarement assis. Ces dernières statuettes, qui sont les plus populaires, même "à l'étranger", représentent l'enfant avec le doigt à la bouche. C'est une attitude qui a connu de nombreuses variantes et qui est héritée des représentations pharaoniques.

Une partie de ce groupe iconographique représente Harpocrate adolescent appuyé contre un pilier profilé à ces extrémités. Ce type semble être apparu au 3e siècle av. J.-C.. Ces pièces représentent Harpocrate parallèlement vêtu ou non d'un chi-

REINACH 1910,. 66-67: des maisons à proximité du temple.

43 DUNAND 1979, n° 178 = 32866.

44 DUNAND 1990, n° 266.

45 WIJNGAARDEN 1958, n° 23, pl. 5, n° 24, pl. 6, n° 36, pl. 8.

46 WEBER 1914, n° 123, pl. 12.

47 WEBER 1914, n° 77-78, pl. 6, n° 89, pl. 9, n° 115, pl. 11, n° 131-132, pl. 12.

48 HENNE 1925, 19, p. 98-99, MICHALOWSKI etc. 1938, 2-22, MICHALOWSKI etc. 1950, 109; DUNAND 1990, n° 110, 270, Le Caire JE 68794.

49 BALLET 1988, p. 406.

50 DUNAND 1990, n° 163.

51 MALAISE 1991.

ton. Leur fabrication a perduré avec une intensité diverse jusqu'au 3e siècle apr. J.-C.. Ce type de représentation qui présente plusieurs variantes était répandu dans tout le pays. Bien que la provenance soit souvent inconnue, on peut constater que ces statuettes étaient utilisées partout en Égypte par une population hellénisée. Les lieux d'acquisition ou les sites de fouilles archéologiques donnés sont Alexandrie[52], Memphis[53], Krokodilopolis[54], Karanis[55], Sunufar[56], Hawara[57], Antinoë[58] et peut-être Thèbes[59].

Les terres cuites du type Harpocrate à la corne d'abondance appuyé contre un pilier et portant le chiton, sont parfois composées d'un faucon portant une double couronne, perché au sommet du pilier. Nous allons nous arrêter sur ce groupe.

Harpocrate debout à côté d'un faucon couronné et dressé sur un pilier

La figure du dieu appuyé contre une colonne ou un pilier[60]

52 BRECCIA 1930, n° 257, II. p. 55, pl. XVI/5 – Chatby, pl. XVIII/70, 72-73, XIX/75, 77, GR 7616, VOGT 1924, pl. X, VIII/6. Musée d'Alexandrie M. 7607, 7591.

53 HORNBOSTEL etc. 1973, n° 296.

54 BAYER-NIEMEIER 1988, n° 90.

55 ALLEN 1985, n° 52, p. 356-7.

56 DUNAND 1979, n° 193, pl. LXXXI = JE 62961 = CG 26987.

57 CG 26989.

58 Même en deux variantes, voir DUNAND 1990, p. 75, n° 146; p. 78, n° 157.

59 WEBER 1914, n° 115 "wahrscheinlich".

60 Par ex. les terres cuites représentant Harpocrate au pot.

61 Harpocrate appuyé contre un tronc d'arbre/cep de vigne (par ex. EDGAR 1904, pl. 23, CG 32061) semble avoir été inventé plus tard, sous l'influence d'Apollon hellénistique en dehors de l'Égypte (voir MERCADÉ 1960, 174-176), et a ensuite été introduit en Égypte dans le répertoire des statuettes en bronze.

62 Musée National d'Athènes, un enfant appuyé contre un pilier: RÜHFEL 1984, fig. 95, p. 227; BRECCIA 1930, n° 104-105, pl. 1-2/Q.

63 WINTER 1903, 237/3-4, 238/3, 242/5-6, 243/3, 8, 244/8, etc., SINN 1977, pl. 18, n° 51: Béotie, le commencement du 3e s. av. J.-C., UHLENBROCK 1990, 50-51 (Athènes)

est une innovation dans la statuaire égyptienne[61]. Dans l'art hellénistique, ce thème est bien connu dans la production de Praxitèle[62]. Il a été très vite adapté par les coroplathes[63]. Ainsi, probablement par l'intermédiaire de la Béotie[64] l'adolescent appuyé contre un pilier est aussi apparu très tôt dans les terres cuites alexandrines. L'application de cette composition sur les représentations d'Harpocrate[65] était alors une tâche sculpturale bien connue, facilement réalisable de point de vue technique, mais qui mettait en même temps en harmonie le dieu représenté et le goût hellène. La rigueur de la pose traditionnelle et de la marche majestueuse a été adoucie par la position de la jambe gauche avancée et repliée. En même temps, le pilier crée un environnement qui fait référence au monde hellénistique, puisque les socles hauts rappelaient les statues des dieux posés sur des socles dans les temples et autour de ces derniers[66]. Cette pose pourrait être inspirée indirectement d'un modèle: par exemple le Dionysos de Madrid, celui de Varèse, la composition de Hermès et Dionysos avec les satyres, et encore Apollon sous forme de Lykeios qui étaient aussi appuyés contre une colonne. Rappelons qu'Apollon était identifié à Horus. Cette même composition d'Harpocrate appuyé sur le pilier peut encore être rapprochée des représenta-

[64] Auparavant dite Tanagra voir WINTER 1903, 237-8, 242. Pour les connections internationales des ateliers, voir HAUSMANN 1984, 288-289.

[65] Le support se limite seulement à une partie des compositions d'Harpocrate à la corne d'abondance, on le voit pourtant relativement souvent parmi les terres cuites. Il peut supporter le poids du dieu ou semble être un élément qui comble l'espace, de manière similaire aux amphores disposées quelques fois à cet endroit (WEBER 1914, pl. 11, n° 114, DUNAND 1990, 78, n° 153, BESQUE 1992, pl. 67, n° c - E.362 (Harpocrate au pot) et à la composition de ses prêtres auprès de l'autel portant des offrandes (BAYER-NIEMEIER 1988, pl. 21, n° 106-111 (prêtre tenant la statue d'Harpocarate).

[66] Par ex. WALKER-HIGGS 2000, 254, n° IV.35, p. 256-7, n° IV.36, le faucon Horus dans le sanctuaire perché sur un pilier par ex. sur la tasse de Stabia, WALKER-HIGGS 2001, 284. fig. 9,5. – dans un style égyptien traditionnel par ex. KAMAL 1905, 148; voir par ex. ERNST 2001.

tions d'Eros qui était aussi en rapport étroit avec le dieu[67].

La représentation du dieu appuyé contre un pilier, combinée avec une statuette de dieu ou de déesse remonte peut-être aussi aux œuvres de Praxitèle; ce type de représentation pourrait notamment être attribué à l'effet de l'Artémis de Larnaka, créée probablement par son fils[68]. Le socle sur les terres cuites hellénistiques était volontiers façonné en forme d'hermé ou il tenait souvent la statuette de Déméter ou d'Eros[69]. L'idée d'un support avec la représentation d'un dieu sur les terres cuites en elle-même n'est donc pas unique dans les œuvres de la coroplathie hellénistique et religieuse. Ce thème était répandu dans «le monde entier». Son application dans l'art égyptien est pourtant un phénomène rare. Les parallèles iconographiques les plus proches sont les représentations des prêtres tenant l'étendard des dieux. Leur fonction est toutefois différente.

A côté d'Horus l'enfant, se trouve souvent une autre figure du même dieu – tradition connue aussi en Grèce – sous une forme plus ancienne du faucon. Celui-ci est couronné du *pschent* et se dresse au centre du sommet du pilier sur la statuette provenant du cimetière Hadra (n° 1). Sur les terres cuites fabriquées plus tard (n° 2-34), le faucon est relégué vers l'extérieur à cause de la position changée de la corne d'abondance.

La plus ancienne terre cuite représentant Harpocrate à la corne d'abondance appuyé contre un pilier sur lequel se trouve un faucon (n° 1) provient d'Alexandrie, comme je l'ai mentionné plus haut, et plus exactement de la partie Manara du cimetière d'Hadra.

67 RIZZO 1932, passim, BIEBER 1961, 16-23.

68 BIEBER 1961, 21.

69 Par ex. HERMITAGE 1976, n° 171, p. 46, avec Aphrodite archaïque sur la colonne, 3e s. av. J.-C.; SCHMIDT 1994, n° 154: Aphrodite d'Asie mineure avec son idole archaïque – 1e s. av. J.-C.; n° 168, Aphrodite anadyomène appuyée contre une figure de Priape ithyphallique de Smyrne, 1e s. apr. J.-C.; n° 180, comme Aphrodite d'Asie Mineure tenant le petit Eros en dessous duquel se trouve la statue archaïsante d'Aphrodite placée sur un socle haut, 2e moitié du 2e s. av. J.-C.

Cette partie du cimetière est daté sur la base de pièces de monnaie de Ptolémée I à Ptolémée VI. L'origine de ce type de terre cuite remonte donc au plus tard au milieu du 2e siècle av. J.-C.[70] A. Adriani limite cette datation – également sur la base des pièces de monnaies – au milieu du 3e siècle pour la partie de Manara[71]. Dans le traitement du manteau de cette statuette, quelques phénomènes caractéristiques peuvent être soulignés: les plis étroits moulent parfaitement le corps, forment un arc et suivent le mouvement du corps. Cette draperie est une caractéristique de la statuaire d'Alexandrie au milieu du 3e siècle[72]. Etant donné la présence de cet élément, cette statuette de terre cuite peut être datée du milieu du 3e siècle, selon toute probabilité. La forme étirée et mince ainsi que légèrement arquée de la corne d'abondance ne ressemble pas à celle des autres terres cuites du groupe, mais se rapproche de celle de quelques oinochoés[73]. La corne d'abondance est également allongée sur un fragment de terre cuite du Louvre[74] dont il ne reste que le buste d'Harpocrate. La position du bras gauche du dieu autorise de penser que ce fragment aurait pu appartenir à une composition du type d'Harpocrate à la corne d'abondance appuyé contre un pilier, mais sans le faucon. La légère rotation du buste rappelle la pose mentionnée. La petite grappe de raisin au sommet de la corne d'abondance est surmontée d'une forme de pyramidion et de petites grenades disposées de la même manière que celle que l'on voit sur les pièces de monnaie de l'époque de la reine Bérénice II[75].

70 KLEINER 1984, 33-34.

71 ADRIANI 1940-1950, 3.

72 cf. BAYER-NIEMEIER 1988, 20-21, note 116, 118; HIMMELMANN 1983, 42.

73 THOMPSON 1973, 32-34, Cat. n° 1, p. 125-126, pl. I, II/1. -II. avec l'inscription dédié à la tyché d'Arsinoé agathé, n° 16, p. 130, pl. VII., n° 25, p. 132-133, pl. IX., n° 34, p. 136-137, pl. XIV – Philopator, vers le milieu, n° 62, p. 144-145, pl. XXI, - Philopator vers la fin, n° 87, p. 153-154, pl. XXXI. – Philopator – voir THOMPSON 1965, 39-44, pl. 15.

74 DUNAND 1990, 80, n° 159 = collection égyptienne, E 29768.

75 THOMPSON 1973, 33.

Le visage triangulaire du dieu ressemble en revanche aux portraits de Ptolémée V.

Ces deux statuettes correspondent à deux variantes d'un thème iconographique datant probablement du milieu du 3e siècle. La variante sans faucon semble être beaucoup plus populaire. Des changements progressifs sont perceptibles dans la forme du vêtement et dans le travail de la tête. Par exemple, une statuette sortie d'un atelier d'Athribis[76] représentant le dieu vêtu du chiton et d'une draperie tombant de l'épaule gauche et enroulée autour de la taille, date probablement de l'époque de Ptolémée VI. La draperie s'arrête au pied du pilier. Du pli central du vêtement sort un phallus surnaturel qui atteint le socle de la pièce. Une autre statuette de ce type, peut-être originaire d'Alexandrie, a dû être créée à la fin du 2e siècle ou au début du 1e siècle av. J.-C.[77]. Des offrandes alimentaires complètent la composition.

Une œuvre du Fayoum[78] (n° 16) atteste non seulement de la survie du faucon dans la composition – phénomène qui n'était pas généralisé – mais aussi de l'intégration dans la représentation de différents aspects du dieu. La statuette représente Harpocrate avec l'index dirigé vers la bouche, tenant une corne d'abondance de la main gauche et appuyé contre un pilier. Mais le dieu porte un chiton long avec des plis qui, au niveau du buste, sont orientés vers le pilier. Au niveau de la taille, le vêtement forme un grand pli triangulaire divisé en petits plis obliques, tandis que les plis sont verticaux dans le bas. Sur le côté extérieur du pilier profilé aux extrémités se trouve un faucon couronné du pschent. Le dieu

[76] MYSLIWIEC 1992, 17, fig. 2. = MYSLIWIEC 1991, 44, fig. 29.

[77] FISCHER 1994, 87, 270, n° 594, pl. 61 = VOGT 1924, p. 10, n° 103, pl. 18,6

[78] Le Caire CG 43263 = JE 42047. Traces de peinture: bordeaux à côté de la corne d'abondance, sur le disque et sur le chapiteau; rose sur la base du pilier. Engobe brun épais jusqu'au milieu du dos. Dim.: 21 cm. Argile brun foncé. Provenance selon le registre, Kiman Farès. Inédite. La description est faite à partir de mes notes réalisées en 1986 avec la permission de M. Qadry†, à qui je voudrais ici exprimer toute ma gratitude.

également couronné du pschent, complété à l'arrière d'un disque de soleil, porte la tresse de l'enfance. Le socle n'est pas modelé. Une statuette de Copenhague[79] d'origine inconnue et peut-être un peu plus ancienne, présente un plissé du vêtement proche de la manière de l'époque romaine. L'intérêt de cette pièce réside dans l'uræus saillant de la corne d'abondance et dans le pyramidion (à moins que ce ne soit un faucon miniature) placé à son sommet. Une statuette du Caire[80] – sans le faucon – présente des plis tout à fait similaires à la série impériale, avec un traitement beaucoup plus souple du drapé. Toutefois, de nombreux détails montrent qu'elle appartient à une autre série de production, tout comme la terre cuite d'Harpocrate appuyé[81].

Ce type peu fréquent de statuettes de terres cuites hellénistiques est suivi à l'époque romaine par un groupe qui est produit en grande série. Le nombre d'exemplaires qui datent du tournant du millénaire témoigne de la popularité de ces statuettes. En dépit de certaines différences, elles ont un archétype commun. Le modèle qui a servi à la fabrication des terres cuites les plus anciennes pourrait être une statue romaine, probablement utilisée lors des grandes processions ou dressée dans un temple. Tenant compte du fait que le thème remonte sans aucun doute au milieu du 3e siècle av. J.-C., le nouveau modèle a pu être une version actualisée d'une statue ancienne, similaire au modèle de la terre cuite de Copenhague illustré ci-dessus[82]. De ce modèle original, il existait plusieurs moules ou plusieurs générations de moules dont la durée d'utilisation pour la fabrication des statuettes a pu varier en fonction de différents critères. En conséquence, il est difficile de connaître avec précision la durée de vie d'un groupe de statuettes issues de ces moules, et à fortiori de dater celles-ci.

[79] MOGENSEN 1930, 43, n° A. 273, pl. XLIII = SCHMIDT 1911, 68, pl. XIII, fig. 32.

[80] DUNAND 1979, 225, n° 194, pl. LXXXII.

[81] DUNAND 1979, 225, n° 193, pl. LXXXIII. trouvée à Sunufar et datée probablement du 1e siècle apr. J-C.

[82] MOGENSEN 1930, 43, n° A. 273, pl. XLIII = SCHMIDT 1911, 68, pl. XIII, fig. 32 (note 79).

Le motif iconographique le plus important de cette série de notre point de vue est le faucon royal à nouveau représenté sur le pilier. Il se dresse sur le côté gauche du pilier, à côté du bras du dieu tenant la corne d'abondance. Ainsi la présence de l'animal semble être explicative, équivalent avec les autres attributs. Le faucon devait avoir son importance dans la composition générale puisque ce type a été fabriqué pendant plusieurs générations. Le groupe sans faucon n'est réapparu qu'après le milieu du 2e siècle apr. J.-C. Ces figures d'Harpocrate sont vêtues d'un long chiton moulant le corps, avec des plis divisés en de multiples petits plis saillants au niveau de l'entrejambe, et elles ne montrent pas le phallus caractéristique des statuettes d'Harpocrates-au-pot. Le vêtement est transparent et le creux du nombril est facilement discernable. Cette manière de traiter le vêtement typique du style romain est également connue sur des stèles funéraires.

La tête du dieu est ronde, les yeux sont profondément enfoncés, le nez est écrasé, les joues sont charnues et séparées du nez et des yeux par des traits gravés. La tête est encadrée par les cheveux arrangés symétriquement en demi-cercles et en volutes, puis couronnée du pschent. Quelques statuettes portent la tresse d'enfance.

Les modifications apportées aux statuettes peuvent être consécutives à la création de nouveaux moules, mais peuvent aussi être le résultat de retouches apportées à des moules usés ou de retouches faites sur des statuettes inachevées. De nombreux détails révèlent ces changements: par ex. l'exécution de la couronne et du ruban ou leur omission, la taille du pilier, les détails du visage, le degré d'inclinaison de la tête, le sens des plis, leur nombre, etc.

La statuette du Musée de Brooklyn (n° 14, Pl. 19) semble refléter une forte influence de l'hellénisme tardif, tandis que celles d'Alexandrie (n° 6), du Caire (n° 17 et 20, Pl. 18/2-3), du Louvre (n° 33), de Budapest (n° 15) et du British Museum (n° 29, Pl. 20) illustrées ci-dessous appartiennent sans doute aux générations suivantes. En observant des cadres des yeux fort retouchés et plus étroits, la ligne arquée des sourcils, les lèvres supérieurs importantes en forme d'arc double, on peut envisager pour cette série un moule original ou un patrice commun fabriqué au tournant du millénaire. Dans le cas de la pièce de Copenhague (n° 21) on peut observer une dualité intéressante: sur un des yeux, les paupières et le dessous de l'œil forment des bosses indépendantes l'une de l'autre comme sur les pièces des générations suivantes (n° 10, 13, 14, 18, 19, 23, 27, 28, 32), tandis que le cadre de l'autre œil est formé par un bourrelet et l'œil lui-même est une dépression. On peut peut-être en conclure que la retouche du moule usé (peut-être un moment abandonné) a été effectuée dans l'intervalle entre ces deux tendances stylistiques. La forme des yeux et le traitement du drapé reflètent encore une forte influence de l'hellénisme tardif. Le style des pièces les plus tardives, comme celles de Liebighaus (n° 24 et 25), suit déjà une mode postérieure au milieu du 2e siècle apr. J.-C. Les détails de deux fragments de tête (n° 11, 34) sont très proches des statuettes de Hambourg (n° 27) et de Copenhague (n° 21) et appartiennent éventuellement à une génération suivante, tout en étant indépendantes l'une de l'autre. Leurs moules étaient usés et estompés. Le fragment de Berlin (n° 11) a été en particulier fort retouché. Ce type de tête est caractéristique de la composition d'Harpocrate avec un faucon juché sur le pilier. Mais en considérant la documentation actuellement disponible dans les publications, la possibilité qu'il ait appartenu à une composition sans faucon ne peut être exclue.

Sur le pilier de la pièce du British Museum (n° 29) figure une inscription qui a été apposée avant la cuisson. Ces signes hiéroglyphiques peuvent être compris comme des paroles prononcées par Harpocrate (Götterrede): *dj-j mrrj / mr kj*[83] *sšw.* "*Je donne l'amour de l'écriture*" ou "*Je donne, que l'autre (c'est-à-dire le peuple/les étrangers*[84]*) aime (aussi) l'écriture*" (Pl. 21). Cette lecture

83 Voir Erman-Grapow, Wb. V. 110 pour la manière d'écrire le *kj* ainsi à l'époque
gréco-romaine. Je voudrais remercier ici Peter Kaplony pour avoir attiré mon attention sur cette possibilité de lire et d'interpréter les signes.

84 Au lieu du pluriel, le mot kj peut signifier dans ce context "*le peuple, les étrangers*". HANNIG 1995, 879.

peut expliciter un aspect d'Harpocrate si l'on considère que l'inscription décrit soit le thème représenté, soit les qualités du dieu, soit qu'elle précise l'objectif de la figuration, soit encore qu'elle véhicule une certaine propagande[85]. Quant au type iconographique, on peut voir une connexion lointaine entre cette composition et celle du type "écolier" de la coroplathie hellénistique; et quant au rôle d'Harpocrate mythique et religieux, il est aussi proche de ce thème.

Si cette inscription est envisagée comme un texte grec composé en hiéroglyphes, ce qui ne serait pas sans parallèle[86], la lecture pourrait être la suivante: on peut lire le *dj* (D38 au lieu de D37) = d, le dieu (A40) = i, la charrue (U7) = m, la bouche (D21) = t mal écrit. Pour ce qui est de la lettre suivante, elle est située sur le bord du chiton et sa lecture n'est donc pas tout à fait certaine: la corbeille (V31) = k ou la bouche déformée (D21) = r. Les derniers signes sont deux lignes obliques (Z4) = i, puis *sš* (Y3) = s et *w* (Z3) = o. Ces signes mis l'un à côté de l'autre forment le nom *Démétrios*. Celui-ci pourrait être le nom de la personne qui a produit la pièce ou le nom de son propriétaire. De nombreuses terres cuites portent un nom grec[87]. Généralement le nom du propriétaire ou du donateur était écrit sur la pièce une fois celle-ci achevée. On pourrait donc être en présence du nom d'un producteur ou d'un artisan potier.

Le lieu où travaillait cet éventuel producteur ou artisan est hypothétique. Etant donné le nombre restreint de statuettes dont la provenance est connue et étant donné que la description, ainsi que les dimensions sont peu satisfaisantes, il est difficile d'établir des rapprochements entre les statuettes. On peut penser que il travaillait soit à Alexandrie, soit plus probablement dans le Fayoum. Les exemplaires appartenant à ce groupe sont largement dispersés, ce qui montre la popularité du thème, mais laisse peut-être aussi entendre que

85 GRAINDOR 1939, p. 23-26.

86 GRAINDOR 1939, 26; KUENTZ 1936, 120-123; MALAISE 1994, 373-383.

87 GRAINDOR 1939, 19-22; voir aussi NACHTERGAEL 1994, p. 413-433.

plusieurs ateliers en possédaient des modèles; il est possible que des moules de générations différentes aient été utilisés simultanément dans un même atelier. Si tel était le cas, il est peu probable que l'inscription qui nous intéresse concerne un artisan-créateur. De fait, la statuette n'est pas de première génération. L'inscription était apposée sur la statuette et non pas sur le moule. Peut être l'artisan inscrivait-il son nom sur la figurine comme une marque de fabrique, afin d'authentifier sa production[88]. Dans cette optique, la question du "droit de production" se pose.

On a observé que les pièces provenant de Myrine inscrites aux noms de Diphilos ou Papios avaient une iconographie différente de celle des œuvres traitant d'un thème similaire exécutées par d'autres coroplathes de l'époque (par ex. Ménophylos et Pythodóros)[89]. La situation devait être semblable en Égypte. Les exemplaires trouvés sur le sol égyptien étaient fabriqués pour le marché local. On peut supposer d'une part que les statuettes ont été produites dans de grands ateliers – en fonction de leur apparence uniforme et du matériau de base qui est similaire – où les pèlerins pouvaient les acheter et les emporter vers leurs villes ou villages d'origine; d'autre part les détails iconographiques et stylistiques révèlent l'existence de plusieurs générations parallèles. Il n'est donc pas impossible que différents ateliers égyptiens aient été en possession de moules identiques. On sait peu de chose de l'organisation des ateliers. Ils pouvait être spécialisés ou bien associés à des ateliers de céramique[90] ou de métallurgie[91]. Des recherches complémentaires permettront peut-être un jour de connaître ce Démétrios éventuel et de savoir s'il dirigeait un atelier et si cet atelier était spécialisé ou multidisciplinaire.

[88] Jaimee P. Uhlenbrock, The coroplast and his craft dans: UHLENBROCK 1990, 15.

[89] BURR-THOMPSON 1934, 8.

[90] Par ex à Athribis, voir MYSLIWIEC 1991, 44.

[91] Beryl Barr-Sharrar, Coroplast, Potter and Metalsmith in: UHLENBROCK 1990, 31-36.

Certaines statuettes présentent des variations incomplètes par rapport à l'archétype: le pilier est un simple support[92]. Celles-ci suivent les séries plus proches de l'iconographie originale avec le ruban sur l'épaule, mais le pilier plus étroit ne laisse pas de place pour le faucon. Le décolleté en V du chiton a été remplacé par un arrondi, à l'exception des pièces de Francfort[93] et d'Alexandrie[94] dont les plis sont moins plastiques, la corne d'abondance plate et les fruits en relief. La facture est peu soignée et la qualité artistique est moindre. Les traits du visage sont grossiers, la présentation de la couronne et des cheveux est plus simple ou peu reconnaissable. Bien que leur racine iconographique remonte à un même modèle, toutes ces pièces appartiennent à des séries indépendantes.

La statuette de Madrid[95], qui est proche de celles de Budapest, de Londres et d'Alexandrie a pu être produite à la fin du 1e siècle ou au début du 2e siècle apr. J.-C.. Le tracé linéaire et les détails grossiers des autres pièces sont typiques des générations plus récentes. Ces pièces ont été fabriquées parallèlement aux séries avec un faucon au cours du 2e siècle, mais avant l'ère des Sévères – les détails stylistiques de la terre cuite de Francfort la situe durant cette époque[96]. Ici le vêtement est fortement linéaire, la tête est disproportionnée, le nez est allongé et entouré de deux lignes verticales. Les yeux en amande sont

92 HORNBOSTEL, etc. 1973, 191, n° 296, pl. 132: "graphische Faltenstil mit spitzwinkligen Brechungen" après 200; TÖRÖK 1995, 73-74, n° 79, pl. XLV; LAUMONIER 1921, 118, n° 578, pl. LIII 3; DUNAND 1990, 76, n° 148; SKUPINSKA-LOVSET 1978, 127, n° UT 121, pl. XXII; BAYER-NIEMEIER 1988, 100, n°. 121, pl. 23,3. – 180/200 n. Chr. = KAUFMANN 1913, fig. 37,7, KAUFMANN 1915, pl. 23, n° 168; BRECCIA 1934, 23, n° 56, pl. XIX, 75; Alexandrie, Musée Gréco-Romain. 7616; Ashmolean M., 18872.1033.

93 BAYER-NIEMEIER 1988, 100, n° 121, pl. 23,3. - 180/200 n. Chr. = KAUFMANN 1913, fig. 37,7; KAUFMANN 1915, pl. 23, n° 168.

94 BRECCIA 1934, 23, n° 56, pl. XIX, 75.

95 LAUMONIER 1921, 118, n° 578, pl. LIII.3.

96 BAYER-NIEMEIER 1988, 100, n° 121, pl. 23,3.

cernés. Les cheveux et la couronne sont indissociables, la corne d'abondance est entièrement droite et placée verticalement. Si la partie inférieure n'était pas si étroite, on pourrait prendre cette pièce pour une massue. D'après les détails iconographiques et stylistiques, la production de ces statuettes a commencé un peu plus tard que celle du type « romain » avec le faucon sur le pilier. Le nombre de terres cuites de cette série indique que la popularité de la composition a beaucoup diminué dans le 3e quart du 2e siècle, et que la production a probablement été interrompue peu de temps après sans doute en raison de la détérioration des moules usés et déjà moins courants.

La terre cuite de Bonn[97] a été retrouvée à proximité de Saqqarah, vers Memphis, un autre exemplaire dans les environs d'Alexandrie.[98] Cette répartition connue présente en apparence une grande disparité par rapport aux pièces dont l'iconographie est plus complète. Cependant, les informations que nous en avons sont trop restreintes et les exemplaires moins nombreux ne permettent pas de donner une répartition réelle à cette série.

Interprétation proposée

Les pièces de provenance inconnue peuvent être datées sur la base de leur style entre le 3e siècle av. J.-C. et le 2e siècle apr. J.-C. Le contexte dans lequel ces statuettes ont été créées, le public auquel elles étaient destinées et l'usage réel qui en était fait, de même que les changements que l'on peut observer dans les formes des statuettes ainsi que dans la variété des attributs du dieu pourraient éventuellement expliquer pourquoi le dieu est devenu aussi populaire en Égypte pendant les premiers siècles de l'empire romain.

La multiplicité des attributs et leur association révèlent un arrière-plan très complexe. Du point de vue du symbolisme égyptien, les attributs du dieu sont bien ceux du dieu Harpocrate, qui

[97] HORNBOSTEL etc. 1973, 191, n° 296, pl. 132.

[98] Alexandrie, Musée Gréco-Romain. 7616.

représente Horus-l'enfant, celui qui incarne le nouveau roi, la continuité et la stabilité du pouvoir. La corne d'abondance qui souligne la fertilité est un autre trait de ses caractères qui appartient en revanche au monde grec. Du point de vue stylistique, l'accent est mis sur l'aspect hellène, qui rappelle aussi que les premiers acquéreurs étaient de couches sociales plus élevées que celles des indigènes de chora.

La corne d'abondance était un symbole officiel des Ptolémées qui est ensuite devenu en Égypte une image courante exprimant le symbole général de la fertilité. Dans ce groupe de statuettes, le phallus surdimensionné surgi au-dessous du pli central du chiton qui est une caractéristique des Harpocrates-au-pot et qui est une autre méthode de mettre l'accent sur l'idée de fertilité, n'est que rarement représenté. La corne d'abondance faisait partie intégrante de la politique religieuse des Ptolémées, parce qu'elle symbolisait l'épithète grecque "*euergetés*" des règnes hellènes, symbole qui fut repris par cette dynastie[99].

Les Harpocrates à la corne d'abondance sont presque toujours coiffés du *pschent*[100], et ils sont peu souvent accompagnés par les deux boutons de lotus, des attributs caractéristiques des autres représentations d'Harpocrate[101]. Le sens théologique de la couronne royale[102] a aussi été déterminé par des sources contemporaines: le *pschent* prescrit par le décret de Rosette rappelle le couronnement de Memphis[103]. Les

99 TÖRÖK 1995, 60, DUNAND 1991, 204-205.

100 Avec le disque solaire: BRECCIA 1934, n° 80, pl. 20, avec le bouton de lotus - WEINGAARDEN 1958, n° 18.

101 Par ex. WEBER 1914, n° 114, DUNAND 1979, n° 191, petit, avec le bouton de lotus, par ex. BRECCIA 1930, n° 257, II. n° 69, pl. 17, DUNAND 1990, n° 154-155, p. 78, BAYER-NIEMEIER 1988, pl. 17, n° 89.

102 Kronen, LdA III. 811-13; RÄRG p. 394-95; ABUBAKR 1937, 60-62; KÁKOSY 1983, 57-60.

103 Urk. II. 192; CAPART 1932, 27; voir KÁKOSY 1983, 60, note 34.

Harpocrates à la corne d'abondance sont en général coiffés aussi d'une couronne de fleur qui rappelle les fêtes en relation avec le dieu. La position de la jambe gauche avancée que l'on observe sur ces statuettes munies de la corne d'abondance est moins fréquente dans les autres groupes-types, par exemple celui sans attribut dans la main[104], celui à la massue[105], celui au pot[106] ou bien chez ses serviteurs du culte avec la statuette du dieu[107], avec la situle[108] ou avec la corbeille[109].

Le faucon coiffé du *pschent* est aussi un élément important. Comme Pascale Ballet l'écrit, l'oiseau devait probablement représenter en même temps deux aspects du dieu[110]. La connexion entre le faucon et Horus, comme symbole de la royauté, était claire pour tout individu vivant en Égypte – indigènes ou étranger. En revanche, pour la population enterrée au cimetière de Manara, qui était composée d'Hellènes ou d'individus fortement imprégnés de la culture grecque, l'aspect d'Horus assimilé à Apollon, dont l'oiseau consacré était aussi le faucon, était sans doute privilégié[111]. En outre, le dieu a aussi été représenté appuyé contre un support. On peut se demander alors si l'identification des dieux égyptiens et grecs a exercé une influence sur les coroplathes dans le choix de ce type de terres cuites ou si la multiplicité des types de statuettes résulte d'un développement propre à l'esthétique de l'époque gréco-romaine.

L'identification d'Horus/Harpocrate avec Apollon a pu contribuer à l'introduction d'un pilier dans l'iconographie d'Harpocrate[112]

[104] Par ex. PERDRIZET 1921, n° 120, 122.

[105] FISCHER 1994, n° = KAUFMANN 1913, n° 100.

[106] WEBER 1914, n° 110, WEINGAARDEN 1958, n° 21,

[107] DUNAND 1979, n° 334, 336.

[108] PERDRIZET 1921, n° 117-199.

[109] PERDRIZET 1921, n° 358.

[110] BALLET 1988, p. 457-458.

[111] Voir Apollo dans LIMC.

[112] La colonne est aussi connue - PERDRIZET 1921, n° 121. - Fayoum; BESQUE 1992,111, pl. 69, n° E.371; WEINGAARDEN 1958, 6, pl. V, n° 17;

inspirée par la colonne agyieus[113]. En prenant en considération des représentations d'Harpocrate avec un pot rond[114], une amphore[115], une galette ronde[116] ou avec l'autel[117] du côté gauche, qui représentaient certainement des offrandes destinées au dieu, on peut penser que cet Harpocrate était la figure centrale d'une scène d'offrande. La statuette ferait alors allusion à un des rites du culte dévolu au dieu.

Les représentations peuvent être rapprochées de scènes qui ornent des oinochoés ptolémaïques en faïence, sur lesquelles les reines Arsinoé II et Bérénice II, souvent vêtues de la robe d'Isis, font des offrandes tandis qu'elles sont figurées tenant la corne d'abondance dans la main gauche, à côté de la colonne d'Apollon[118]. Ces vases et les phialés décorés de l'effigie des Ptolémées étaient des cadeaux de Nouvel An. Les oinochoés servaient à faire une libation à cette occasion[119]. Il est probable que les scènes figurées sur les vases représentaient ou condensaient des évènements de la fête du Nouvel An à Alexandrie, moment où la reine faisait des offrandes pour le bien-être du pays et de son peuple. Le roi, en tant d'Horus, devait probablement aussi faire des offrandes à cette occasion, ce qui pouvait également être représen-

DUNAND 1990, 77, n° 152, p. 78, n° 155, p. 81, n° 164, et à côté des Harpocrates-au-pot, en position debout.

113 Tandis que le support d'Harpocrate en dehors de l'Égypte est représenté en général par le cep de vigne qui était en connection avec Dionysos, et qui donne l'impression d'une sorte d'amalgame des statues d'Apollon au carquois et de Eros ailé. La composition syncrétique était répandue parmi les statuettes en bronze (REINACH 1909, vol. II. 481, 483; vol. IV. 297-299; MALAISE 1991a). Un moule provenant d'Égypte se trouve à EDGAR 1904, pl. 23, CG 32061.

114 DUNAND 1979, n° 194.

115 DUNAND 1979, n° 193

116 EDGAR 1904, CG 32062, pl. 24.

117 EDGAR 1904, CG 32062, pl. 24, MOGENSEN 1930, A. 273.

118 THOMPSON 1973, passim.

119 THOMPSON 1973, 71-75, 122.

té sur des objets. Ce ne serait pas exceptionnel, étant donné que la dynastie ptolémaïque était connue pour sa théâtralité, les rois étaient souvent figurés comme des dieux[120]. Le rôle du culte du roi et de la reine mis en parallèle n'est pas un cas unique non plus à l'époque hellénistique – cela était également prescrit par décret[121]. D'autres sources attestent du rôle intermédiaire joué par le roi et par les offrandes pour obtenir l'abondance dans le pays[122]. Bien que la force du culte royal des époques pharaoniques se soit atténuée, la continuité était néanmoins réelle, comme on peut le noter pour certains empereurs romains [123]. Plusieurs éléments ont été empruntés aux *vota publica*.

La composition montrant Harpocrate appuyé contre un pilier peut être interprétée à plusieures reprises comme une réelle scène d'offrande. Depuis le 2e siècle av. J-C., des terres cuites représentent quelque fois le dieu à côté des offrandes; un moule du Caire plus tardif[124], qui représente le dieu à la corne d'abondance debout entre un pilier et un autel, semble faire ressortir exactement ce fait. Dans la scène d'offrande, il n'y a pas de vase à libation dont la représentation aurait empêché de représenter le doigt du dieu devant la bouche. L'existence de l'offrande est aussi signalée par la présence de l'autel. Le support en forme de pilier et sa forme rappellent à la fois les supports utilisés pour porter les statues des dieux lors des processions égyptiennes et la base des statues des dieux hellènes et romains dans les temenos publics[125]. La question se pose de savoir si

[120] Joan B. Connelly, Hellenistic Alexandria dans: UHLENBROCK 1990, 89-92.

[121] Par ex. III. Ptol. Euergetes, décret de Canope 34, lignes 40-41, V. Ptolemaios Epiphanes – décret de Rosette lignes 37-41, 46-51

[122] Par ex. POSENER 1960, 47-61, des reliefs des temples d'Edfou, Dendera, Philae, etc.

[123] THOMPSON 1973, 122-124, ALFÖLDI 1964-1965, p. 53-87, voir aussi Apul. Metamorph. XI. 17,2-3.

[124] EDGAR 1904, CG 32062

[125] Par ex. TÓTH 1978, 585-595.

lors de la formation de cette composition on a voulu représenter une offrande harpocratique générale ou bien une offrande rattachée à un évènement ou à un rite spécifique.

La connexion étroite entre la composition et le culte royal se manifeste non seulement par la double présence du *pschent* sur le dieu et sur l'oiseau, mais aussi par la représentation du faucon[126]. Bien que les pharaons Égyptiens aient souvent été représentés avec l'oiseau et que le faucon sur le pilier apparaisse également comme l'une des formes du nom d'Horus[127], le faucon était rarement représenté sur les terres cuites. Ici, il semble y avoir une connexion entre le faucon et la corne d'abondance. La seule exception est une terre cuite romaine datant probablement du 2e siècle où le faucon se dresse sur une colonne, à côté d'un Harpocrate jambes croisées, sans la corne d'abondance[128]. Le faucon se dresse sur le pilier dans une autre composition (sur le pied cylindrique d'une lampe) où Harpocrate à la corne d'abondance est assis sur un cygne[129]. Le faucon semble être aussi sur la corne d'abondance d'un statuette de Berlin représentant Harpocrate assis sur un lotus, tandis qu'à côté de la corne d'abondance se dresse un uræus. Il n'est pas impossible que les trois éléments – la corne d'abondance, l'uræus et le faucon – se trouvent réunis dans la main d'un autre Harpocrate appuyé contre un pilier[130]. Cette intéressante association donnerait: le roi (= le faucon) qui est lui-même un dieu (= uræus) prend soin du bien-être du pays (= corne d'abondance) par sa forme d'Horus l'enfant (= doigt dans la

126 Voir TÖRÖK 1995, 73: «The presence of the falcon ... stresses the identity of the childgod with Horus, heir of Osiris, and associates him not only with the act of offering ... but also with the concept of divin sonship and also hints at an identification of the terrestrial ruler with Horus. The association with the dynasty is also indicated through the cornucopia, a frequent attribute of Ptolemaic kings and queens».

127 *Ḥr ḥr w3ḏ.f*, voir KEES 1915, 66-67, 75.

128 PERDRIZET 1921, n° 121, pl. XXVI, à gauche, en bas (Fayoum).

129 BALLET 1988, 454, note 3: Alexandrie, 9442(?)

130 MOGENSEN 1930, A. 273.

bouche, tresse d'enfance, *pschent*) à l'aide des offrandes aux dieux (= pilier).

Nous savons que les Ptolémées favorisaient Sérapis dans la fonction de règne. Si l'aspect «*euergetés*» des Ptolémées était en connexion avec Harpocrate sur les terres cuites plutôt que Sérapis, cela devait évoquer un cas très particulier en rapport avec le prince héritier[131].

De fait, les statuettes d'Harpocrate reproduisent parfois les traits des Ptolémées[132]. Parmi les statues de Ptolémée V, une tête calcaire avec une tresse d'enfance couronnée d'un *pschent* est bien connue[133], de même parmi d'autres représentations, notamment les terres cuites[134]. Des exemples antérieurs à Ptolémée V peuvent être cités. Un camée représente Ptolémée IV en tant qu'enfant-roi avec les attributs d'Harpocrate[135]. Il est aussi représenté jeune sur une statue monumentale, coiffé du *pschent,* mais sans la tresse de l'enfance[136]. Un texte bilingue atteste que Ptolémée IV a fait construire un petit temple pour Harpocrate dans le temenos du Serapeum[137]. Plus tôt encore, des éléments de l'iconographie d'Harpocrate apparaissent sur les monuments de Ptolémée III: sur un camée le roi est représenté enfant, coiffé du *pschent,* probablement au début de son règne[138]. Le rapprochement de la fonction du roi et l'enfant pouvait exprimer l'espoir du renouvellement continuel de la vie et de la joie dans le pays. Si dans la composition précitée le pharaon et l'enfant Horus à la corne d'abondance sont unis pendant l'offrande on peut associer cette

[131] Je voudrais ici remercier à Prof. László Török qui a attiré mon attention sur cette contradiction.

[132] TÖRÖK 1995, p. 60-61; VOLLENWEIDER 1984, 363-377.

[133] KYRIELEIS 1975, 54-56, 134-136, fig. E.1; WALKER-HIGGS 2001, n° 14, 51.

[134] KYRIELEIS 1975, 54, pl. 42/5 = BAYER-NIEMEIER 1988, n° 167.

[135] VOLLENWEIDER 1984, p. 375, n° 25, pl. LXVI/1. = VOLLENWEIDER 1995, n° 71.

[136] KREIKENBOM 1992, 123-124, I16.

[137] RONCHI 1974, 153; FRASER 1972, vol. I. 261; vol. II. 412, note 569.

[138] VOLLENWEIDER 1995, n° 62.

image du jeune roi qui vient d'être installé sur le trône à la prospérité du pays[139]. Il est possible que les rois aient été représentés à l'occasion de leur couronnement, et que les coroplathes aient été inspirés par de telles évènements. En ce cas une variante en terre cuite de la statue de culte de Ptolémée V réalisée à cette occasion a peut-être également été créée[140]. Certaines statuettes de bronze représentent Harpocrate coiffé du *pschent* avec un faucon à côté de la jambe. M. Malaise a émis la possibilité qu'elles évoquent aussi le couronnement [141].

Mais existait-il vraiment une occasion lors du couronnement ou lors d'un événement qui y était lié au cours duquel le faucon – vivant ou sous forme de statue – était réellement présent? Les textes liturgiques d'Edfou citent les cérémonies de «*la fête du couronnement*» le 1e du mois Tybi[142]. Au cours de celles-ci, en tant qu'âmes de Hiéraconpolis et de Bouto, les prêtres portaient un pavillon contenant la statue du roi avec le faucon dressé sur le *serekh,* du sanctuaire au balcon de couronnement. L'union symbolique du roi et du faucon réalisée dans le temple, ils paraissaient à nouveau devant le peuple. Puis le faucon vivant et la statue du roi étaient placés dans un naos et recevaient les offrandes avant d'être une nouvelle fois présentés au peuple. Les mêmes cérémonies avaient lieu pour commémorer le couronnement le 1e du mois Thot. On connaît aussi les rites d'Athribis[143]. Le papyrus du Musée de Brooklyn n° 47.218.50[144] raconte la fête de l'anniversaire du couronnement dont la date fictive et dogmatique était celle du Nouvel An[145]. Les oiseaux

[139] Voir DOBROVITS 1937, p. 67-122, qui met l'accent sur la représentation du dieu ayant dépassé l'enfance, mais n'étant pas encore adulte.

[140] TÖRÖK 1995, 58.

[141] MALAISE 1993, 147-158; cf. MALAISE 1991, 13-35; S. SCHOTT 1968, 54-65.

[142] ALLIOT 1949-1954, 309-433, 565sqq.

[143] ALLIOT 1949-1954, 216, 284, note 10.

[144] GOYON 1972.

[145] Königskrönung, LdÄ III. 531-33, RÄRG 395-400.

(*ʿʿn.t*) vivants d'Horus jouaient là encore un rôle important, de même qu'au couronnement du Ptolémée XII Aulètes.[146] En conséquence, il n'est pas impossible que la composition ait été inventée pour un couronnement ou pour l'anniversaire de celui-ci et qu'elle se rapporte simultanément aux cérémonies et au roi identifié à Horus. La représentation de la couronne de fleurs fait également penser à une occasion solennelle[147] Cette interprétation peut expliquer l'existence du type au faucon ainsi que la composition similaire sans faucon: elles peuvent représenter deux moments différents d'un évènement. Puisque le roi en place exécutait plusieurs rites sans la présence du faucon, la dernière version devait être plus répandue. L'invention de la composition du cimetière d'Hadra (n° 1) peut ainsi être située entre le règne de Ptolémée III et celui de Ptolémée V, souverains qui non seulement diffusaient leur culte royal hellène mais voulaient aussi gagner des fidèles de la religion égyptienne en soutenant le culte royal égyptien.

La composition devait être en même temps adaptée au culte funéraire, puisqu'une nouvelle période, une nouvelle vie débutait avec le couronnement et le Nouvel An, de la même manière que la vie nouvelle commençait après la mort, lorsque le défunt régnait après avoir été assimilé au dieu Osiris, le père d'Horus qui a lui-même "*fait vivre son corps (sc. d'Osiris) dans le château de l'or*"[148]. Le roi était alors en ce cas Horus *m-ḫt krs jt-f*, celui qui assurait l'ensevelissement d'Osiris/du défunt après la mort.

Le transfert du symbolisme royal au domaine de l'Au-delà est une tendance générale dans l'histoire de la religion égypti-

146 BERGMAN 1968, p. 92-120. On ne sait pas si, sous les Ptolémées, au cours des fêtes du couronnement ou de son anniversaire, le roi ne pouvait pas porter un vêtement avec des décorations faites de plumes de faucon, comme cela se faisait pendant le Nouvel Empire. Cf. ALDRED 1969.

147 PW IV, 1636-43, XI. 1588-1607; BAUS 1940, 19-23; BLECH 1982, 302-311. cf. Apuleius Metamorphoses IX.

148 CAUVILLE 1980, 60. et 50, note 30.

enne. Cette tendance peut également être observée dans les arts mineurs à l'époque saïte par exemple dans les amulettes en forme de couronnes de Haute ou de Basse Égypte. Durant l'époque hellénistique, les fidèles de la déesse Isis avaient une préférence pour les motifs de la couronne[149] qui sont également fréquents dans les décors des tombes gréco-romaines. A Athribis, célèbre pour son culte du faucon, le banquet funéraire se déroulait en présence des oiseaux d'Horus[150], probablement pas sans antécédents

Après que l'Égypte ait été conquise par l'Empereur Auguste, le gouvernement égyptien et en conséquence la vie et le mode de réflexion du peuple ont été complètement modifiés. C'est ce qui explique que la composition de l'archétype de ce groupe diffère en plusieurs aspects de celle d'Hadra (n° 1), encore qu'elle était l'antécédent. Elle ne reflète pas le gout individuel d'un maître ou le résultat d'une copie mal réussite, car beaucoup de coroplathes ont continué à modeler cette composition pendant une longue période avec des variations mineures. Les déviations de la composition d'Hadra sont les conséquences naturelles des siècles passés, mais la nouvelle formulation visuelle ne peut pas être que le reflet intellectuel de l'époque pour ce qui cette nouvelle composition d'Harpocrate est formulée. Sa création et son utilisation se rattachent probablement à une exigence générale. Si ces statuettes n'avaient pas pu être vendues, elles n'auraient pas été produites en série. Etant donné qu'elles étaient destinées au marché intérieur[151], la référence à des éléments spécifiques à l'Égypte doit être retenue pour ce type.

149 KÁKOSY 1983, 57-60, WESSETZKY 1961, 43-44.

150 PARLASCA 1970, 173-198.

151 Dans d'autres compositions il était connu, par ex. le faucon près des jambes d'Harpocrate (cf. les articles de Malaise) ou le faucon sur la colonne à côté d' Apolloniskos à Delos, dans le livre des inventaires du Serapieion C (ROUSSEL 1916, 221, 278 = MARCADÉ 1960, 174, notes 4 et 7 = BRUNEAU 1970, 168.) – à 156/155 apr. J.-C.

Comme la tête de la statuette d'Hadra n'a pas subsisté, les comparaisons ne peuvent s'appuyer que sur les autres parties de la statuette. La structure de la composition n'est pas modifiée. Au lieu des plis animés de la statuette hellénistique on trouve ici un chiton statique en symétrie rigide et un décolleté en V qui s'est transformé en arc vers la fin de la production de ce groupe (n° 3-33). Le dieu immobile paraît distant des fidèles, ce qui accentue son caractère divin. Le fait que l'extrémité du phallus soit en général cachée, à l'inverse des statuettes d'Harpocrate-au-pot appuyé contre une colonne produites à la même époque, n'est pas un hasard[152]. Cela suggère la fertilité d'un manière générale, exprimée seulement par la corne d'abondance, qui pouvait se manifester dans n'importe quel aspect de la vie et qui devait déterminer le destin individuel.

La différence dans la position du faucon sur le pilier est importante par rapport à l'origine : par sa position à côté de la corne d'abondance son rôle complémentaire est clair. Il est peu probable que dans l'Égypte romanisée le faucon surmonté du *pschent* ait été une allusion au faucon d'Apollon. Le faucon était un motif familier en Égypte en tant que symbole royal. Il est également peu probable qu'il ait fait référence à un évènement spécial vécu par le roi en personne, car les empereurs n'étaient pas souvent dans le pays et leur "couronnement" y était réalisé seulement de manière virtuelle[153]. La

[152] La représentation du phallus surmesuré parle de la contamination des deux groupes, p. ex. on peut le voir à une autre occasion, avec la corne d'abondance sur le pilier, mais nu, et un phallus énorme entres les jambes (DUNAND 1990, n° 154, p. 78), comme aussi le fait, qu'Harpocrate ait été représenté parmi les terres cuites appuyé contre le pilier et tenant le pot, p.ex. sans le chiton et le faucon, il a la corne d'abondance près de sa jambe gauche par terre et il tient un pot rond au-dessus du pilier (DUNAND 1990, no. 136, p. 72), etc..

[153] La seule exception est Vespasien, qui était considéré comme un dieu par les Alexandrins quand il a été proclamé roi. Cet évènement s'est pourtant déroulé sans aucune connexion avec l'intronisation traditionnelle. Cf. MILNE 1924, 29.

composition peut toutefois faire allusion à une forme particulière du culte royal. Parmi les statues escortées pendant les processions, on trouve des statues d'Harpocrate à la corne d'abondance appuyé contre un pilier, sur lesquelles le dieu n'est pas vêtu du chiton et sur lesquelles le faucon est absent[154]. La situation est identique pour les statues du culte. Cependant, dans le cas de la statue de Ras el-Soda trouvée dans le sanctuaire romain d'Isis, on suppose qu'un faucon était perché sur le bras[155]. Mais là encore des problèmes se posent. Il est tout naturel que le fils d'Isis regina ait été lié à l'idée de la royauté, représenté avec le faucon, mais la composition de cette statue reproduit un autre aspect d'Harpocrate, le héros hellène, avec un autre symbolisme que les compositions des séries de terres cuites évoquées ci-dessus.

Un phénomène caractéristique du culte royal de l'époque romaine étaient les *vota publica*, au cours desquels quelques cérémonies des rites du Nouvel An ptolémaïque ont également perduré. Les cérémonies consistaient en offrandes faites dans l'ensemble de l'Empire romain pour le bien-être de l'Empereur. Elles étaient complétées par des éléments du mystère d'Isis, comme les offrandes à la déesse et parmi eux la consecration des statu(ett)es[156]. Là on trouve également la statuette de son fils, Harpocrate. Si l'on accepte l'idée que le type hellénistique traité est lié aux fêtes du couronnement, sous quelque forme que ce soit, alors on peut également supposer que la statuette romaine qui possède la même structure en constitue le prolongement. En conséquence, la production des compositions romaines doit peut-être sa prospérité aux fêtes des *vota publica*.

Comme l'objectif des *vota publica* n'était pas seulement d'assurer le bien-être de l'Empereur, mais plutôt de lui exprimer loyauté et dévouement, le dieu-fils Harpocrate devait représenter

154 BARTA 1968, 75-78, GYŐRY 1983, 53-72.

155 ADRIANI 1935-39, 140-142.

156 cf. ALFÖLDI 1937.

dans ce contexte l'enfant d'Isis regina et d'Osiris-Serapis omnipotens, le prince héritier qui pouvait transmettre la vigueur de l'innocence des enfants et la promesse d'un futur bienheureux. Cette image a pu être complétée par celle d'un Harpocrate, en tant que héros de culte qui – si nous acceptons l'interprétation selon laquelle l'inscription sur la terre cuite de Londres est l'explication égyptien de la personne représentée – avait pour but de propager la science de l'écriture, et par l'écriture transmettait aux fidèles toutes les branches de la science: l'astrologie, la médicine, etc. L'idée prend racine à l'époque des pharaons et est aussi connue à l'époque hellénistique en Égypte. Horus est en effet mentionné au Moyen Empire en tant qu'écolier-type ayant de nombreux talents. Ce trait de sa personnalité n'a pas été oublié au cours de l'époque romaine, sur la base de l'histoire démotique de Siuserré[157], et aussi dans l'arétalogie de Chalcis du 2e siècle apr. J.-C. il est présenté comme le détenteur des sciences et de la culture[158].

On remarque ici aussi une dualité des idées – la statuette de Karanis a été trouvée à proximité du sanctuaire sans contexte précis, tandis qu'à Sunufar elle a été mise au jour dans le cimetière. Mais on exigeait une protection renforcée et des connaissances étendues pour pouvoir assurer la vie dans l'Au-delà, donc les mêmes fonctions pouvaient très bien servir dans les deux mondes.

Les multiples facettes de ces terres cuites permettaient plusieurs usages, comme jadis les amulettes: accompagner les défunts dans leur voyage vers la salle des deux Maat où Osiris les jugeait traditionnellement, être offertes aux dieux dans les temples où les statues des empereurs se trouvaient parmi celles des dieux-*synnaoi* ou être disposées dans les maisons avec d'autres dieux lares afin de bénéficier de leur aide sur terre. La statuette assurait ainsi partout à l'individu la possibilité d'une vie

157 KAPLONY 1966.
158 HARDER 1944, 7-18.

meilleure par le biais des traditions anciennes réinterprétées, grâce au prince héritier érudit et aux dieux cosmocrators qui le soutiennent. On comprend maintenant aisément pourquoi Harpocrate était aussi populaire partout en Égypte.

Hedvig Győry
Musée des Beaux-Art
Budapest

Addition

Pendant la Symposium sur les dieux enfants ptolémaïques et romains en Égypte à Mainz (2002) j'ai montré l'image de la terre cuite du British Museum avec l'inscription qui se trouve sur le pilier. Quelques semaines plus tard, j'ai reçu une lettre de H-W. Fischer-Elfert.

Notiz zur Inschrift auf der Terrakotte BM von H.-W. Fischer-Elfert:

Die kurze vertikale Inschrift auf dem Pfeiler lautet wie folgt:
dj-j mr-k zẖ.w "ich lasse dich die Schriften / das Schreiben lieben"
Der Wortlaut dieses kurzes Satzes gibt sich nach Ausweis des Suffixes der 1. Sing. als von einem Gott gesprochen zu erkennen. Angesichts der Darstellung der Statuette, dürfte vermutlich Harpocrates sich dahinter verbergen.

Wer dem aber auch immer sei, vielleicht noch bemerkenswerter könnte die Provenienz deises Vers(teils!) sein, findet sie sich doch in genau dieser Gestalt in Kap. III c der aus dem Mittleren

Reich (MR) datierenden Lehre des Chety[1] wieder. Vollständig müßte der Vers lauten:
dj-j mry-k zẖ.w r mwt-k "Ich lasse dich die Schreiben / das Schreiben mehr lieben als deine (eigene) Mutter."

Ob der Komparativ auf der Terrakotte einzig aus Platzgründen ausgefallen worden ist oder andere Gründe hat, halte ich einstweilen für nicht eruierbar. Inwieweit der (gesamte) Text dieser Lehre in hellenistisch-römischer Zeit noch verfügbar und in gelehrten Priesterkreisen noch bekannt war, bleibt ebenfalls bis zum Auftauchen eindeutig in die gleiche Epoche zu datierenden Handschriften eine müßige Frage. Angesichts der aus Ptolemäischer Zeit stammender Fragmente der Lehre des Amenemhets I. [2] (gleichfalls MR) ist mit dieser Möglichkeit, nicht zuletzt auch in Anbetracht des antiquarischen Interesses an alten Texten, unbedingt zu rechnen.

[1] Zählung nach W. Helck, Die Lehre des *Dw3-Ḫtjj*, Teil I, Wiesbaden 1970, S. 28.

[2] S. Pap. Berlin P. 23045 bei G. Burkard - H.W.. Fischer-Elfert, Verzeichnis der Orientalischen Handschriften in Deutschland, XIX.4, Stuttgart 1994, Nr. 177. J.F. Quack wird demnächst weitere Fragmente dieser Handschrift in der ZÄS publizieren.

ANNEX

		Prove-nance	Hauteur	Matériel	Publication
1.	Alexandrie 25745	Hadra / Manara	8 cm		ADRIANI 1940-1950, 19. pl. V.
2.	Alexandrie 22580	Chatby	12,4 cm		BRECCIA 1930, 55, no. 257, pl. XVI/5.
3.	Alex andrie 5134	inconnue	22 cm		BRECCA 1934, 23, no. 58, pl. XVIII, 70.
4.	Alexandrie 5131 (avec la t te d'Isis)	inconnue	22 cm		BRECCIA 1934, 23, no. 57. pl. XVIII. 72.
6.	Alexandrie 7616	inconnue	18 cm		BRECCIA 1934, no. 57. pl. XIX, 77.
7.	Alexandrie 7606	inconnue	19 cm		BRECCIA 1934, 23, no. 57. pl. XVIII. 73
8.	Alexandrie 7591	inconnue	17,3 cm		BALLET 1988, 69, no. 94.
9.	Ann Arbor, Kelsey Museum 9956	Karanis	20 cm	brun rouge (5 YR 5/4)	ALLEN 1985, 356-357, no. 52.
10.	Berlin, Ägyptisches Museum 8795	inconnue	20, 4 cm	brun du cuir	WEBER 1914, 95, no. 118, pl. 11.
11. ?	Berlin, Ägyptisches Museum 16384.	Batn Harit	5 cm	brun de terre	WEBER 1914, no. 70, pl. 6.
12.	Brooklyn, Museum of Fine Art 37.1622E Pl. 19/1-4.	Saqqarah	19,2 cm	brun	(BAYER-NIEMEIER 1988, no. 117, ALLEN 1985, 357, no. 52.)*
13.	Bruxelles, MRAH E. 3808 Pl. 18/6	inconnue	19,1 cm	brun	(ALLEN 1985, 357, no. 52.)*
14.	Bruxelles, MRAH E. 3932 Pl. 18/5	inconnue	18,8 cm	brun	(ALLEN 1985, 357, no. 52.)*

15.	Budapest, Musée des Beaux-Arts T. 478	inconnue	20,7 cm	brun de brique	TÖRÖK 1995, 73, no. 78, pl. IX.
16.	Cairo CG 43263 = JE 42047	Kiman Far s	21 cm	brun sombre	–
17.	Cairo, CG 26987 Pl. 18/3	Sunufar	20 cm	brun rouge	DUNAND 1979, 227, no. 197. (ALLEN , 357, no. 52.)
18.	Cairo, CG 26988 Pl. 18/1	inconnue	20 cm		DUNAND 1979, 229, no. 213.
19.	Cairo, CG 26989 = JE 19805 Pl. 18/4	Hawara, Labyrinth	18 cm		DUNAND 1979, 229, no. 214. (ALLEN p. 357, no. 52.)*
20.	Cairo, JE 42264 Pl. 18/2	inconnue	20,5 cm		DUNAND 1979, 227, no. 199.
21.	Coppenhague, NY Carlsberg AE.I.N. 385	inconnue	16,5 cm		MOGENSEN 1930, 43, no. A. 277, pl. XLIII. = SCHMIDT 1911, 68, pl. XII, fig. 31.
22.	Coppenhague, NY Carlsberg AE.I.N. 387	inconnue	20 cm		MOGENSEN 1930, 43, no. A. 278 = SCHMIDT 1911, 68, pl. XII, fig. 30.
23.	Frankfurt, Liebighaus 2400.1117	inconnue	19,8 cm	brun rouge, sombre	BAYER-NIEMEIER 1988, 99, no. 117, pl. 22,5. = KAUFMANN 1913, 51, fig. 28,17 = KAUFMANN 1915, 52, pl. 18, no. 116.
24.	Frankfurt, Liebighaus 2400.1118	inconnue	20,2 cm	Brun rouge, sombre	BAYER-NIEMEIER 1988, 100, no. 118, pl. 22,6.
25.	Frankfurt, Liebighaus 2400.1119	inconnue	20,9 cm	Brun rouge, sombre	BAYER-NIEMEIER 1988, 100, no. 119, pl. 23,1.

26.	Frankfurt, Liebighaus 2400.1120. (pi ce inférieure non-inhérente)	inconnue	21,2 cm	Brun rouge, sombre	BAYER-NIEMEIER 1988, 100, no. 120, pl. 23,2.
27.	Hamburg, Museum für Kunst und Gewerbe 1989.342	inconnue	19,6 cm	brun rouge	EWIGLEBEN 1991, 47, no. 18 = HORNBOSTEL etc. 1977, 167, no. 144.
28.	Hildesheim, Pelizäus Museum 717/42	inconnue	18,3 cm	brun rouge	(BAYER-NIEMEIER 1988, no. 117, ALLEN 1985, 357, no. 52.)*
29.	London, British Museum 27333/96.3-30.2. Pl. 20-21	inconnue	18,7 cm	brun rouge	(ALLEN 1985, 357, no. 52.)*
30.	Lyon, Musée des Beaux-Arts 195-73	Koptos (fouilles de Reinach en 1911.)	20,7 cm		BALLET 1988, 70, no. 97.
31.	Paris, Louvre AF 7489	inconnue	15,7 cm	brun rouge (P 29)	DUNAND 1997, no. 149.
32.	Paris, Louvre E.20852	inconnue	20,6 cm	brun rouge (P 33)	DUNAND 1997, no. 151.
33.	Paris, Louvre E. 20.642	inconnue	18,4 cm	brun jaune (P 75)	DUNAND 1997, no. 150.
34.	Paris, Louvre E 20742 A	inconnue	4,5 cm	brun rouge (P 49)	DUNAND 1997, no. 323.

RELIGIONSGESCHICHTLICHES ZUR PANNONISCHEN ISIS

Bei einer Rettungskampagne der Mitteldanubischen Regionalen Altertumsverwaltung im Jahre 4025 identifizieren die Archäologen in einer insula des einstigen Budapester Stadtbezirks "Elisabethstadt" unter den Trümmern eines vierstöckigen Hauses eine unverwechselbare ghanäische Kulttrommel.[1]

Dank der Fundsituation stellen die Fachleute fest, daß ihr einstiger Besitzer der Ägyptologe Hu-Ba-I war. Prosopographischen Forschungen zufolge ist der Name schon im Namensmaterial des chasarischen Kaganats dokumentiert, es kann aber eine indirekte Verbindung nach Afrika nicht ausgeschlossen werden. Die Trommel kann ziemlich genau datiert werden, weil der Epigraph der Expedition auf dem zerfallenden Zeitungspapier, in das die Trommel eingepackt war, Nachrichten über das Nationaltheater, die Europäische Union und die Inflation entziffern kann. Man vermutet, daß die Trommel um die Wende zum 3. Jahrtausend an ihren Fundort gelangt ist. Der Leiter der Ausgrabung verkündet der Öffentlichkeit: Endlich sind uns exakte Beweisstücke in die Hände gelangt, die unsere schon lange gehegte Vermutung bestätigen, daß ein guter Teil der Budapester Bevölkerung in jener Zeit Anhänger einer animistischen Religion war, deren rituelle Tänze von afrikanischer Trommelmusik begleitet wurden. Der ghanäische Ursprung der Trommel gestattet weitere Schlußfolgerungen bezüglich des Kultes. Es kann davon ausgegangen werden, daß die Budapester Männer und Frauen in Strohröckchen an den Zeremonien teilnahmen, wenngleich die Spezialisten noch darüber streiten, ob die Strohröckchen der Budapester Animisten als

[1] Für die Publikation wurde der Vortragsstil weitgehend beibehalten.

Beweis orgiastischer Fruchtbarkeitskulte, oder gegenteilig - im Spiegel paleometeorologischer Forschungen – als Beweis asketischer Selbstzüchtigung betrachtet werden müssen. Abgesehen von diesen Detailfragen jedoch stimmen die Forscher darin überein, daß die Trommel aus Ghana die Ausbreitung des afrikanischen Animismus in der genannten Region anzeigt. Auf Grund der ghanäischen Trommel beschäftigen sich die Archäologen mit einer erneuten Bestimmung mehrerer, zuvor als Konservenbüchse identifizierten Funde, welche nach dem heutigen Stand der Wissenschaft als Kindertrommeln gedeutet werden. Mit Hilfe dieser, zunächst als Konservenbüchse mißgedeuteten Funde arbeiten die Forscher an der Topographie der Ausbreitung der animistischen Kulte. Der mitteldanubische Oberpriester des ghanäischen Kultes könnte der einstige Ägyptologe gewesen sein, denn von ihm stammt das bisher größte und schönste derartige Denkmal. Zur Aufbewahrung dieses bedeutenden Fundes hat die lokale Selbstverwaltung ein Museum errichtet, welches von Tausenden von Touristen und Schulkindern besucht wird. Diese epochale Erkenntnis und der bedeutende Fund gelangten in die Zeitungen, in Volksbildungsprogramme und natürlich auch in die Schulbücher. Auch ich habe auf diese Weise davon erfahren.

Nun zum Ernst der Sache: Die hermeneutische Schwierigkeit und Frage der Interpretation besteht darin, ob wir die vorliegende Kultur von den Funden her interpretieren müssen, oder die Funde von der Kultur her. Sollen wir die zum sehr großen Teil unbekannte "antike Kultur" von den immer zufälligen Funden her interpolieren oder in den Hintergrund einer vermeintlich bekannten Kultur einige Mosaiksteinchen einfügen? Die Frage sollte nicht mit einem voreiligen, irenischen "sowohl als auch" gelöst werden, weil damit das Dilemma des hermeneutischen Zirkels eliminiert würde. Wenn wir die Frage nicht ernst nehmen, werden wir sie kaum authentisch beantworten können.

Meiner Ansicht nach besteht gegenwärtig unsere Aufgabe nicht darin, auf Grund der zutage tretenden lateinischen Inschriften, ägyptisierenden Denkmäler oder tatsächlich aus Ägypten stammenden Gegenstände darzustellen, wo überall weitere Kultstätten des vermutlichen Isis-Kultes gewesen sein könnten, in der

Annahme, daß es an den Fundorten Isis-Verehrer gab. Umgekehrt! Wir müssen in der nur mosaikartig bekannten, sogenannten pannonischen Kultur den Ort der ägyptisierenden oder tatsächlich ägyptischen Gegenstände finden und das Vorhandensein des ägyptischen Geistes beweisen können. Es ist eine Tautologie, wenn wir den Fundort auf der Karte als einen neuen Kultort einzeichnen, denn dies beweist nicht mehr, als daß an dieser Stelle der vorliegende Gegenstand gefunden wurde (was ohnehin schon mit der Entdeckung bekannt ist), der Fundgegenstand muß aber auch etwas über das Vorhandensein der lebendigen ägyptischen Frömmigkeit aussagen. Sollten die Funde nicht von selbst in das Bild passen, dürfen wir sie trotz ihres physischen Vorhandenseins nicht dorthinein pressen, wo sie nicht hingehören. Kann man glauben, daß die in Begram gefundene, Bronzestatue des Sarapis[2] (der Stolz des Museums von Kabul) das Beweisstück der Anwesenheit ägyptischer Kulte in Afghanistan sein soll? Wer glaubte im Ernst, daß der Osiris-Altar von Balatonszabadi von einem Osiris-Kult in Balatonszabadi zeugt? Aber warum sollte es sich mit dem Fragment der *jmȝḫw*-Stele von Aquincum anders verhalten? Oder was sollen wir von dem, in Báta/Komitat Tolna gefundenen echten ägyptischen Uschebti[3] aus der Kaiserzeit halten, wo doch im Mutterland mit dem Uschebti des Pa-scheri-en-Ptah aus der Zeit Kleopatras VII. diese Fundgruppe längst ausgestorben war?

Da sich bisher im Zusammenhang mit der vermuteten, pannonischen ägyptischen Religiosität nur Archäologen und Ägyptologen geäußert haben, möchte ich meinerseits zwei weitere, bisher noch nicht herangezogene Disziplinen zur Sprache kommen lassen: die historische Religionssoziologie und die Religionsgeschichte.

Wenngleich die Religionssoziologie als historische Disziplin nichts Neues ist, scheint es, als ob mit Rodney Starks Buch "The Rise of Christianity. A Sociologist Reconsiders History" (Princeton

[2] ARSLAN 1997, 545
[3] LAKATOS 1961, 10, Nr. 5.

1996) wieder etwas in Bewegung geraten wäre. Dieses Werk macht den Versuch, das Wachstum der christlichen Gemeinschaften in den ersten 300 Jahren in einer Aufsplitterung nach Jahrzehnten in Prozenten zu erfassen. Er untersucht nicht nur die gesellschaftliche Schichtung der Kultteilnehmer näher, sondern auch die fördernden und behindernden Faktoren (Judenmission, Rolle der Frauen, Ehebräuche, Geburtenregelung, Epidemien, die Stadt als Faktor, Märtyrer). Die Gläubigen rücken in lebendige Nähe. Die Sache ist aus dem Grund interessant, weil sich der Religionswechsel, obwohl er in gesellschaftlichem Ausmaß geschah und statistisch erfaßt werden kann, nicht auf Grund des Prinzips "cuius regio eius religio", sondern in der Folge einer persönlichen Entscheidung vollzog. Wie ist dies zu erklären? Diejenigen, die sich einer neuen religiösen Bewegung anschließen, rekrutieren sich im allgemeinen aus den zuvor religiös Inaktiven. So waren auch die frisch bekehrten Christen in ihrer Mehrheit ehemals inaktiv, denn wenn sie eine Zugehörigkeit gehabt hätten, wären sie nicht konvertiert. Da sie jedoch religiös inaktiv waren, kann man davon ausgehen, daß ihre Entscheidung nicht von theologischen Erwägungen bestimmt wurde. Auf Grund moderner Forschungen wissen wir, daß solche Menschen eine neue Religion annehmen, die der alten Religion gegenüber skeptisch geworden sind. Dies ist in Hinblick auf die gesellschaftliche Position typisch für die relativ Privilegierten, jedoch nicht für die Oberschicht, die viel zu verlieren hätte. Bewegungen gesellschaftlicher Gruppen rekrutieren sich über die interpersonellen Verbindungen von Gruppenmitgliedern und Außenstehenden. Eine der Voraussetzungen für das Annehmen einer neuen Religion ist die Auflösung der Normenkonformität – wenn sich die interpersonellen Verbindungen lockern, ist der Wechsel leichter. Dies ist in Städten, wo sich die Menschen nicht kennen, eher der Fall.

Da Claude S. Fischer in seiner Theorie von der 'Subkultur der Urbanisation' behauptet, daß mit zunehmenden urbanen

[4] STARK 1997, 156.

Charakter eines Ortes auch die Unkonventionalitätsrate[4] steigt und sich das Milieu für den Religionswechsel vergünstigt, schreibt Stark der städtischen Verbreitung der neuen Religion eine hervorgehobene Rolle zu. Deshalb untersuchte er die Entwicklung des Christentums[5] in den zweiundzwanzig mehr als 40.000 Einwohner zählenden Städten des Römischen Reiches. Er stellt unter anderem fest, daß die Verbreitung der neuen Religion in Relation zu der Entfernung der betreffenden Stadt von Jerusalem steht und in zweiter Linie auch durch die Abhängigkeit von der Reichshauptstadt Rom beeinflußt wurde.

Dies alles ist deshalb interessant für uns, weil es bezüglich der Isis-Religion zwar schon prosopographische[6] Studien bzw. gender studies[7], jedoch noch kein Herangehen in der oben genann-ten Weise gibt. Jetzt aber hat Stark in der Person von Tim Hegedus[8] einen Anhänger gefunden, der die Beziehung der 44 größten Städte der Ptolemäerzeit und der frühen Kaiserzeit zur Isis-Religion untersucht hat. Primär legt er die Entfernung von Alexandria, sekundär die Entfernung von Rom zugrunde. In seinen Resultaten ist er kritisch genug, um nicht die Ausbreitung in mecha-nischer Weise in Abhängigkeit zu der Entfernung oder der Bevölkerungsgröße dieser Städte zu setzen, wenngleich sich herausgestellt hat, daß der Isis-Kult bis zur Geburt Christi nicht weiter als 1000 Meilen über Alexandria hinaus gekommen ist. Dies sollte erst im Römischen Reich geschehen. Die Rolle der Stadt Rom ist unbestreitbar, aber die Bedeutung des viel kleineren Rhodos oder Delos in der Frühzeit muß unbedingt beachtet werden. Es lohnt auf jeden Fall, sich mit seinen Ergebnissen auseinanderzusetzen, an dieser Stelle jedoch möchte ich nur hervorheben, worauf ich durch ihn gestoßen bin. Wenngleich Hegedus bei der Auswahl der Städte mit

[4] STARK 1997, 156.

[5] In Anlehnung an CHANDLER - FOX 1974.

[6] MORA 1990.

[7] HEYOB 1975.

[8] HEGEDUS 1998, 161-178.

Isis-Kult eine gewisse Großzügigkeit walten ließ, indem er alle jene Städte einbezieht, in denen irgendein archäologisches Fundstück des Isis-Kultes vorliegt (ich würde strenger vorgehen und nur die Städte mit einem Isis-Tempel untersuchen und nicht jeden archäologischen Fund für einen Beweis des einstigen Isis-Kultes betrachten), ist dennoch auffällig, daß bei 32% dieser Städte keine Spuren einer ägyptischen Religiosität zu finden sind. Dies ist in jedem Fall eine signifikante Zahl und ich bin daher vorsichtig mit der Behauptung, daß die sogenannte Isis-Religion in der Römerzeit allgemein verbreitet gewesen ist.

Es kann näher untersucht werden, in welcher der mehr als 30.000 Einwohner zählenden Städte kein ägyptischer Kult zu finden war: im pontischen Amasia, Apamea am Ufer des Orontes, Nisisbis, Damascus, Ekbatana, Edessa, Jerusalem, in Arabien in Rayy, Marib und Zafar (Jemen), im sizilianischen Messana und im hispanischen Cadiz und Sevilla. Wenn wir nun von den leicht verständlichen hispanischen Städten und dem schwer erklärbaren Messana absehen, können wir das Fehlen des Isis-Kultes in der östlichen Hälfte der antiken Oikoumene konstatieren. Besonders überrascht, daß es in dem so nahe bei Ägypten gelegenen Jerusalem keine Spuren der ägyptischen Religion gibt, wo doch hier ein stark hellenisiertes Judentum wohnte, dessen Verbindung zu Alexandria allgemein bekannt ist und wo nach dem zweiten jüdischen Krieg sogar die Stadt den Juden verboten war, dennoch reichte die Isis-Propaganda der Antoninier und Severer nicht aus, diese Religion hier zu verwurzeln. Nicht weniger auffällig ist die Isis-abweisende Attitüde von Damascus und sonstiger Städte des Seleukidenreiches sowie der südarabischen Städte. Ähnlich auffällig ist schließlich die "Isis-Losigkeit" des pontischen Amasia, das insgesamt nur 160 km von Sinope, der angeblichen Heimat der ersten Kultstatue des Sarapis entfernt liegt.

Wenn wir nun die Landkarte ohnehin vor uns haben, fällt uns auf, daß der tatsächliche Isis-Kult auf ein relativ enges geographisches Gebiet im Becken des Mediterraneum begrenzt ist: auf das Nordufer Afrikas, das Westufer Kleinasiens, Griechenland und Italien. Alles andere ist sehr sporadisch, wenn nicht verdächtig - so auch der pannonische Kult.

Die bisher nicht mit dem notwendigen Gewicht gestellte Frage, warum die Isis-Religion sich nirgendwo östlich vom östlichen Ufer des Mittelmeeres ausgebreitet hat, ist schwer zu beantworten. Weiterführen könnte vielleicht die Beantwortung der Frage, warum sie dort Anklang fand, wo sie sich ausgebreitet hat. Dies würde der weiteren Forschung einen wichtigen Anstoß geben. Vielleicht war die dortige Kultur nicht genügend hellenisiert bzw. später romanisiert, um diese griechisch-römische (jedoch nicht ägyptische) Göttin annehmen zu können. Aber damit habe ich schon vorausgegriffen und auch nur einen Aspekt der Antwort berührt.

Wollen wir nun das religionsgeschichtliche Herangehen betrachten![9] S. Morenz[10] beschreibt die ägyptische Religion bei ihrer Typisierung als Kultreligion. Dieser Kult ist ein Staatskult, ihr Agent ist der. Pharao der als Vermittler für den Staat, ja sogar für die Ordnung des Kosmos auftritt, jedoch nicht für das Heil[11] des Einzelnen.

Lange Zeit bedeutete dies kein Problem, da in die Stammes- und Volksreligionen und damit in den Heilszustand, in welchen man nicht hineingetauft werden kann, jedes Glied der Gemeinschaft hineingeboren wird. (Beispielsweise kann jemand, dem der Gedanke der Reinkarnation gefällt, nicht Hindu werden, da er nicht in eine Kaste hineingeboren wurde und da man Hindu nur durch Geburt werden kann.) Aufgabe des Kultes ist in diesen Religionen die Aufrechterhaltung des kosmischen Gleichgewichts und die Abwehr der Gefahr des Hineinfallens in das

9 Meine religionsgeschichtlichen Erwägungen führe ich in einem weiteren Zusammenhang in meinem Aufsatz "Über die Ursachen des Sieges des Christentums in Ägypten" aus, welcher in Numen 48, 2001, 81-116 erschien. Hier habe ich nur einige Gedanken herausgehoben.

10 MORENZ 1960, 244 ff.; MORENZ 1964, 19-28.

11 Ich bin mir dessen bewußt, daß die meisten Lexika und Nachschlagewerke keine Definition für "Heil" geben. Diesen Mangel bemerken wir umso empfindlicher, als dazu z.B. im Englischen gar kein adäquates Wort existiert. "Salvation" kann sowohl Heil als auch Erlösung heißen, welche als Zustand bzw. Vorgang einander gegenseitig ausschließen.

alltägliche Chaos.[12]

Bei dem anderen Typ von Religion (wozu die Mysterienreligionen und das Christentum gehören) wird das Individuum in den Zustand des "Unheils" geboren, und es muß individuell den Weg des Heils finden. Die so zum Heil Gelangten bilden nun eine sekundäre, nicht auf Blutsbanden beruhende Geimeinschaft, Gemeinde oder Kirche. (Diese Gemeinde kann jedoch später als ein Pseudo-Ethnikum betrachtet werden, z.B. verus Israel.)

In der ägyptischen Volksreligion ist es der vom Pharao repräsentierte Staat, welcher als Heilsanstalt das allgemeine Heil für die Ägypter (d.h. für die "Welt") garantiert und die *mꜣꜥt* (maat) verwirklicht. Die *mꜣꜥt*, also das vom König heraufgeführte Heil, ist nichts Metaphysisches, Transzendentes, sondern eine normale Lebens-form, in der auch Gerechtigkeit und Solidarität geübt wird. Der Zustand des Unheils verlangt nach Erlösung, aber das Heils bedarf der Erlösung nicht. Die altägyptische ethnische Religion, die an die zentrale Position der *mꜣꜥt* glaubt, und so meint, im Heilszustand zu leben, kennt weder die Erlösung des Menschen, noch die der Welt. Jan Assmann, vielleicht der einzige Ägyptologe, der je das Wort "Heil" in einem ägyptologischen Buch gebraucht hat, hat also vollkommen recht, daß "die Welt nach ägyptischer Vorstellung nicht der *Erlösung*, sondern nur der *Inganghaltung* bedürftig ist. Sie braucht den Herrscher, nicht den Heiland."[13]

Der als *restitutio mundi* vollzogene kosmische Gottesdienst des Pharao begann zu der Zeit problematisch zu werden, als sich die Gesellschaft im Zeitalter des Hellenismus atomisierte und den Menschen durch ihre religiöse Erfahrung nicht mehr bestätigt wurde, daß sie in einen Heilszustand geboren worden sind. Dieser zentralisierte Kult ist nicht mehr geeignet, dem individuellen religiösen Bedürfnis zu genügen, und der Staatskult wiederum gibt dem Individuum dazu keinerlei Möglichkeit. Deshalb ist das Individuum schließlich zur Magie als einzigem Ausweg verdammt.

12 MENSCHING o.J., 58-70.

13 ASSMANN 1990, 222; das heuer erschienene Buch des Autors "Heil und Herrschaft" war in Ungarn noch nicht erreichbar.

An dieser Stelle kann ich im Zusammenhang mit der Isis-Religion zwei unhistorische Fragen nicht umgehen, die auch für Pannonien von Bedeutung sind:

1. *Warum siegte nicht die Isis-Religion über das Christentum oder "historischer" formuliert: Warum war nicht die Isis-Religion, sondern das Christentum diejenige Religion, die auf die religiöse Verarmung, das existentielle Vakuum der mediterranen Völker eine adäquate Antwort geben konnte?*

Das Defizit der europäischen Isis-Religion gegenüber dem Christentum muß man, glaube ich, nicht im Zentrum der Welt dieser Religion suchen, sondern wir werden es in den speziellen Problemen der Mission finden. Wie Jesus Christus *Soter* ist, ähnlich ist auch Isis *soteir' athanate.*[14] Die Grenzen der sogenannten Isis-Mission werden, worauf S. Morenz aufmerksam macht,[15] dadurch gut illustriert, daß der Oberpriester Mithras[16] den Lucius mit dem "aus dem Geheimgemach des Heiligtums" geholten, "mit unverständlichen Schriftzeichen versehenen" Buch (Metam. XI. 22.) katechetisiert. Vor der Profanisierung, "vor der Neugier der Uneingeweihten" wurde das Buch durch "die kurze Schrift einer Sprache, welche in Bildern von allerlei Tieren abgefaßt war, ... gesichert",[17] das heißt, es handelte sich sicher um Hieroglyphen Schrift. Auf einem aus Herculaneum stammenden Fresko sehen wir einen Morgengottesdienst, der dem von Apuleius beschriebenen Gottesdienst in Kenchreä ähnlich ist und bei dem der in weiß gekleidete, auf den Stufen vor dem Heiligtum stehende Isis-Priester mit versteckten Händen die sogenannte Kanope umfaßt, in der das für das Ritual notwendige Nilwasser enthalten ist. Von den beiden links und rechts neben ihm stehenden, ein Sistrum schüttelnden Männern könnte der rechte ägyptischer Herkunft sein, worauf seine

14 Zitat in: GRIFFITH 1975, 238 aus dem ersten Hymnus Isidors 26 (=SEG 8,548).

15 MORENZ 1961, 432-436. Nachdruck MORENZ 1975, 521-526.

16 In der sogenannten Isis-Mission erscheint häufig der Name des Mithras, während in Ägypten Mithras ganz an der Peripherie bleibt. Vgl. G. Lease, Mithra in Egypt, in: PEARSON – GOEHRING 1986, 114-129.

17 Apuleius, Metamorphosen, hrgs. von R. HELM 1956, 345

dunkle Hautfarbe hinweist. Oben auf der Treppe liegen zwei Sphingen, auf dem Geländer und beim Altar vor den Chorsängern schreiten ägyptische Ibisse.

Zum Isiskult haben diese ägyptischen Bestandteile organisch dazugehört; die Heimat, der nationale Charakter sind mit der Religion verwachsen. (Wir können uns vorstellen, welch ein Problem es in den fernergelegenen Teilen des Imperiums darstellte, einen Priester zu finden, der des Lesens von Hieroglyphen kundig war oder Nilwasser zu beschaffen.) Diese sogenannte Isis-Mission war *typologisch gesehen eine universale Religion*, dennoch *konnte sie sich* in ihrem Kult *nicht von der nationalen Bindung loslösen*. Dafür mußte die Mission mit der Zeit teuer bezahlen. Auch das spätere ägyptische Christentum war stark nationalistisch geprägt; dies hat sich nach dem Bruch auf dem Konzil im Jahre 451 gezeigt. Anststatt sich zum universalen Christentums zu entwickeln, wurde es zur Nationalkirche. Auch der ägyptische Nationalheilige Schenute wurde von keiner der christlichen Gemeinschaften angeeignet.

Das auf jüdischen Ursprung zurückreichende Christentum hatte mit denselben Problemen zu tun. Die Heilige Schrift, welche die Prophetien enthielt, war ursprünglich in hebräischer Sprache geschrieben worden - die christliche Mission jedoch hat sich gerade im Interesse der Verbreitung von Anfang an auf die in der alexandrinischen griechischen Tradition wurzelnde LXX gestützt. Die hebräische Heilige Schrift hat weder im Kult, noch in der theologischen Interpretation eine Rolle gespielt. Das Christentum als Gemeinschaft der Neugeborenen fungierte nicht nur typologisch gesehen als universale Religion, sondern es löste sich auch im Kult von seinen *ursprünglichen orientalischen Bindungen*[18] und hat sich damit gegenüber der soteriologisch ähnlichen Isis-Religion in der

[18] Nach Raffaele Pettazzoni ist bei der Mithras-Religion dieselbe *conditio sine qua non* zur Geltung gekommen: Sie erreichte in dem Maße Erfolge, in welchem sie ihre eigenen orientalischen Elemente ("*ihr asiatisches Barbarentum*") verleugnete und sich der Kultur der griechisch-römischen Welt anpaßte. Siehe: PETTAZZONI 1954, 180-192.

Mission einen riesigen Vorteil verschafft.

Wir sehen also, daß auf die in einer gegebenen geschichtlichen Situation auftretende *challenge* (A. Toynbee) zwei Religionen mit ähnlichen Voraussetzungen zwei unterschiedliche *response* geben: die Isis-Religion bezahlt wegen ihres Festhaltens an der heiligen Tradition mit dem Erlahmen der Mission, das Christentum wiederum bezahlt für seine missionarischen Erfolge, wie jede Mission zu allen Zeiten, mit theologischem Synkretismus. Die sich jedoch als inadäquat erweisende Antwort (*response*), welche also auf die Herausforderung (*challenge*) nicht die "richtige" Antwort gab, bedeutete für die betreffende Richtung, namentlich für die Isis-Religion, deren Tod.

2. *Warum konnte die Isis-Religion, die in Europa so erfolgreich zu sein schien, dem Glauben ihres ägyptischen Mutterlandes keine Kraft geben und warum konnte dieses so sehr erschüttert werden, daß nur noch das Christentum jüdischen Ursprungs ihm eine authentische Lösung geben konnte?*

Wir haben oben gesehen, daß das relative Hindernis der Isis-Religion deren national geprägter Kult war. Aber warum erweist sich diese trotz allem siegreiche Religion nicht als ausreichend auf dem alten ägyptische Boden, wo sich ihr "Nationalismus" vermutlich nicht nachteilig auswirkte? Dabei können mehrere Faktoren eine Rolle gespielt haben, zum Beispiel jener, daß Griechisch die Sprache der europäischen Isis-Religion war; dies könnte in Ägypten bereits ein Hindernis für eine angenommene "innere Mission" gewesen sein. Wichtiger als dies jedoch ist, daß in dieser von uns als europäische Isis-Religion bezeichneten Richtung Sarapis eine wichtige Rolle spielte, zeitweise erscheint noch "Kynokephalos" oder Harpokrates, aber mit ihnen ist die Reihe der ägyptischen Götter mehr oder weniger zu Ende. Im Mutterland jedoch ist in den Tempeln von Philae, Kom Ombo, Edfu, Esna, Dendera u. a. trotz jeder Krise die Palette der Götter wesentlich breiter. Die Mythologie ist jedoch nach unserer Überzeugung zweitrangig.

Meines Erachtens war wohl auch hier das entscheidende Problem, daß die ägyptische Religion national war, während der europäische Isis-Kult zum Typ der Universalreligionen gehörte. Das mochte praktische und theologische Fragen gleichermaßen aufwerfen. Es heißt zum Beispiel in einer Inschrift von Esna: "*Lasse nicht einen Asiaten in den Tempel eintreten, sei er alt oder jung!*"[19], und dies bezog sich wahrscheinlich auch auf jeden asiatischen geweihten Isis-*Mysten*. (Das Problem bestand nicht darin, daß dies kein Isis-, sondern ein Chnum-, Neith- und Hekatempel war. Zudem sind die Ägypter auch nicht im heutigen Sinne des Wortes 'in den Tempel gegangen'.) Auch die Annäherung von der anderen Seite kann man sich schwer vorstellen. Hätten die letzten Hüter des alexandrinischen Sarapieions, Helladios als Priester des Zeus – offensichtlich Amons –, Ammonios als Priester des Affengottes – also Thots – und Olympios (die alle Hellenen und neuplatonische Philosophen waren) an den blutigen Opfern des oberägyptischen Kultes teilhaben können? [20]

Es geht hier schließlich um einen strukturellen Unterschied: Der Ägypter, der seine altägyptische Religion ausübte, nahm an einem "*kollektiven* Bestreben"[21] teil, wohingegen der europäische Anhänger der Isis-Religion, der eine persönliche Entscheidung getroffen hat, als *Individuum*[22] Mitglied einer Gemeinde wurde. *Im Imperium ließ sich der Isis-Gläubige, der als Römer, Grieche, Kleinasiat geboren war, in das Mysterium einweihen und gelangte durch Initiationsriten in die Mysterien-*

[19] SAUNERON 1957, 197,20. Zitiert von Morenz 1975, S. 523

[20] Im Roman des Heliodor schlachteten die Meroiten "allerlei Tiere" für Dionysos. "*An die (Altare der) beiden andern brachten sie für Helios ein Viergespann weißer Rosse ... und für Selene ein Paar Stiere*." (Heliodor, Die äthiopischen Abenteuer von Theagenes und Charikleia X. 6., Übers. Horst Gasse, Stuttgart 1972, 275.) Der weise Priester Sisimithres ist jedoch anderer Meinung, die er aber nur insgeheim zu äußern wagt, "*auf Griechisch, um nicht vom Volk verstanden zu werden*", "*...da nach unserem Sittengesetz allein schon die Gebete und Wohlgerüche genügen!*" (ibid. X. 9. S. 278.).

[21] collective endeavor, wie FRANKFURTER 1998, 6 sich ausdrückt, und S. 266: "... adherence to traditional religion was not an individual affair of personal salvation but a community affair".

[22] HEGEDUS 1998, besonders 162.

gemeinschaft, oder sagen wir, den Anachronismus bewußt übernehmend, in die Isis-Kirche. Demgegenüber *stand der ägyptische Fellache schon seit seiner Geburt in dieser Religiosität. Er wurde nicht in die Osiris-Mysteriengemeinschaft eingeweiht, dennoch war er Teil jener universalen Heilsgemeinschaft wie jedes Glied seines Volkes*. Die Aufgabe war nur, diesen *kosmos*, also diese Ordnung, Anordnung, tagtäglich aufrechtzuerhalten und nicht zuzulassen, daß die zerstörerischen chaotischen Kräfte einbrechen. (Dies ist Aufgabe des Kultes). Wenn wir es wie zuvor mit einem Anachronismus ausdrücken, dann ist er in eine Volkskirche hineingeboren. Der *Übergang zwischen beiden Religionsgemeinschaften – vereinfacht gesagt, der oberägyptische Fellache und der römische Isis-Mystes – wird nicht durch den soziologischen Unterschied erschwert, sondern durch den grundsätzlichen Gegensatz in der Heilsauffassung*. Deshalb wäre eine solche hypothetische 'Innere Mission' für Isis in Ägypten von vornherein zum Scheitern verurteilt gewesen.

Was heißt das? Die in Ägypten lebendige Religion (die etwa anderes als die Isis-Religion ist) und die in Griechenland und Italien sich verbreitende Mysterienreligion sind strukturell gesehen zwei verschiedene Realitäten. Soziologisch gesehen gehörten die Isis-Gläubigen nicht in die ägyptische Religion, dies wollten sie sicherlich nicht und konnten es auch nicht. Sie gehörten jedoch zu einer neuen, durch sie gegründeten Gemeinschaft, die sich in Anlehnung an die Isisreligion ein Mini-Ägypten mit Palmen, Nil, Sphingen und Isis schuf. Der ägyptische Fellache sehnte sich entweder nicht göttlicher Erwählung und Erlösung oder wenn er es tat, dann suchte er diese aus Gründen der Tradition nicht bei der, von ihm schon als heidnisch angesehenen Isis, sondern bei Christus. Zwei Religionen, die beide Isis kennen, und dennoch zwei Welten! Aber ist dies so etwas Besonderes? Auch Juden und Christen sprechen von Mose, aber wie verschieden ist der Stellenwert dieses Mose hier und da. Katholiken und Protestanten sprechen von der Jungfrau Maria, aber die heilsgeschichtliche Rolle Marias und ihr Erleben in der konkreten Frömmigkeit sind ganz verschieden. Für Christen und Muslime ist Jesus gleichermaßen heilig, für die einen ist er aber auch ihr Erlöser.

Wenn es nun gelungen ist, einige Probleme der strukturellen Unterschiedes zwischen der ägyptischen Religion in der Römerzeit und des hellenistischen bzw. römischen Isis-Kultes aufzuzeigen, ist es an der Zeit, uns dem pannonischen sogenannten Isis-Kult zuzuwenden. Die Tatsache, daß Pannonien geographisch weit von Ägypten entfernt liegt und es auch bisher noch nicht gelungen ist, ägyptische Elemente in der Bevölkerung aufzuzeigen (diese wurden höchstens mit dem recht verschwommenen Ausdruck "Orientalen" angedeutet), legen nahe, daß hier eine ägyptische Religiosität kaum zu erwarten ist. Merkwürdigerweise hat sich jedoch das Gegenteil davon in der wissenschaftlichen Meinung verbreitet, bei deren Ausformung zwei ausgezeichnet ausgebildete, über großes Faktenwissen verfügende, gewissenhafte Ägyptologen eine entscheidende Rolle gespielt haben. Aladár Dobrovits hat im Jahre 1943 eine Studie mit dem Titel "Az egyiptomi kultuszok emlékei Aquincumban"[23] veröffentlicht und im Jahre 1959, schon nach den Ausgrabungen in Savaria erschien der Artikel von Vilmos Wessetzky in den Acta Archaeologica.[24] Dieser sowie seine zwei Jahre später erschienene Monographie[25] besitzt auch heute noch bestimmende Kraft. Diesen beiden großartigen Menschen bin ich persönlich verpflichtet und schulde ihnen auch dann Dank, wenn meine Meinung in einigen Punkten nicht mit ihrer Meinung vollständig übereinstimmt. Von Professor Dobrovits habe ich als junger Mensch gelernt zu fragen, Professor Wessetzky hat in seinem Buch erstmals darauf hingewiesen, wie fraglich die *cultuelle* oder *culturelle* Interpretation der auf pannonischem Gebiet gefundenen Gegenstände ist und ob diese auf ägyptische Vorstellungen oder aber auf deren interpretatio romana zurückgehen.[26] Beide haben jedoch eindeutig für eine Anwesenheit ägyptischer Religiosität in Pannonien entschieden und sich dabei neben zahlreichen kleineren Funden vor allem auf die Mumien von Aquincum, die Kanne von

[23] DOBROVITS 1943.

[24] WESSETZKY 1959, 265 ff.

[25] WESSETZKY 1961.

[26] WESSETZKY 1965, S. 145-146.

Egyed und insbesondere den im Jahre 1955 ausgegrabenen Tempel von Szombathely gestützt. Vom Standpunkt der ungarischen Ägyptologen blieb, soweit ich es beurteilen kann, das lateinsprachige, epigraphische Material zweitrangig. Wenn wir jetzt das Vorhandensein oder Nichtvorhandensein ägyptischer Religiosität in Pannonien erwägen, müssen wir notwendigerweise das gesamte Material durchgehen und in Bezug auf eine lebendige Beziehung mit Ägypten bzw. dem ägyptischen Geist in die Waag-schale werfen. Ich möchte hier meine Gedanken zum hellenisti-schen bzw. römischen Isis-Kult und die sich darauf beziehenden bekannten Quellen nicht erörtern, sondern nur einige Fragen formulieren, auch nicht einmal in Bezug auf das vollständige archäo-logische Material.

Über die pannonischen Mumien haben in jüngster Vergangenheit auf einem ähnlichen Symposium Forscher gesprochen, die in diesem Thema kompetenter sind als ich. Diese in mumifizierten Form gefundene Leichname sind *für sich selbst genommen* in den Augen des Ägyptologen noch keine Beweisstücke ägyptischer Religiosität; sie zeugen erst in Verbindung mit gegenständlichen Denkmälern des ägyptischen Totenkultes vom osirianischen Glauben des Verstorbenen oder seiner einstigen Umgebung. In diesem Zusammenhang kommt, wenn ich mich recht erinnere, nur die Mumie von Aquincum in Frage, deren Grab eine bemalte Holztafel beigelegt ist;[27] die kleinen Ausmaße des Tafelbildes jedoch, seine Placierung an den Füßen und seine von den Mumienportraits vermutlich abweichende Funktion bedürfen in meinen Augen noch einer weiteren Erklärung. Allein auf Grund dieses Fundes ist die Hypothese der sogenannten pannonischen Isis-Religion nicht haltbar.

Kommen wir zum Tempel: Robert A. Wild hat in seiner Arbeit "The Known Isis-Sarapis Sanctuaries of the Roman Period"[28] die 44, manche archäologisch nur zum Teil bekannten Heiligtümer des Römischen Reiches gesammelt. Wenn wir davon die 8 in Ägypten fungierenden Tempel abziehen, bleiben noch 36 für das gesamte Reich. Jedoch auch deren Identifizierung als Isis-Heiligtümer ist nicht

[27] PÓCZY 1966, 272 ff.
[28] ANRW II. 17/4, 1984, 1740-1851.

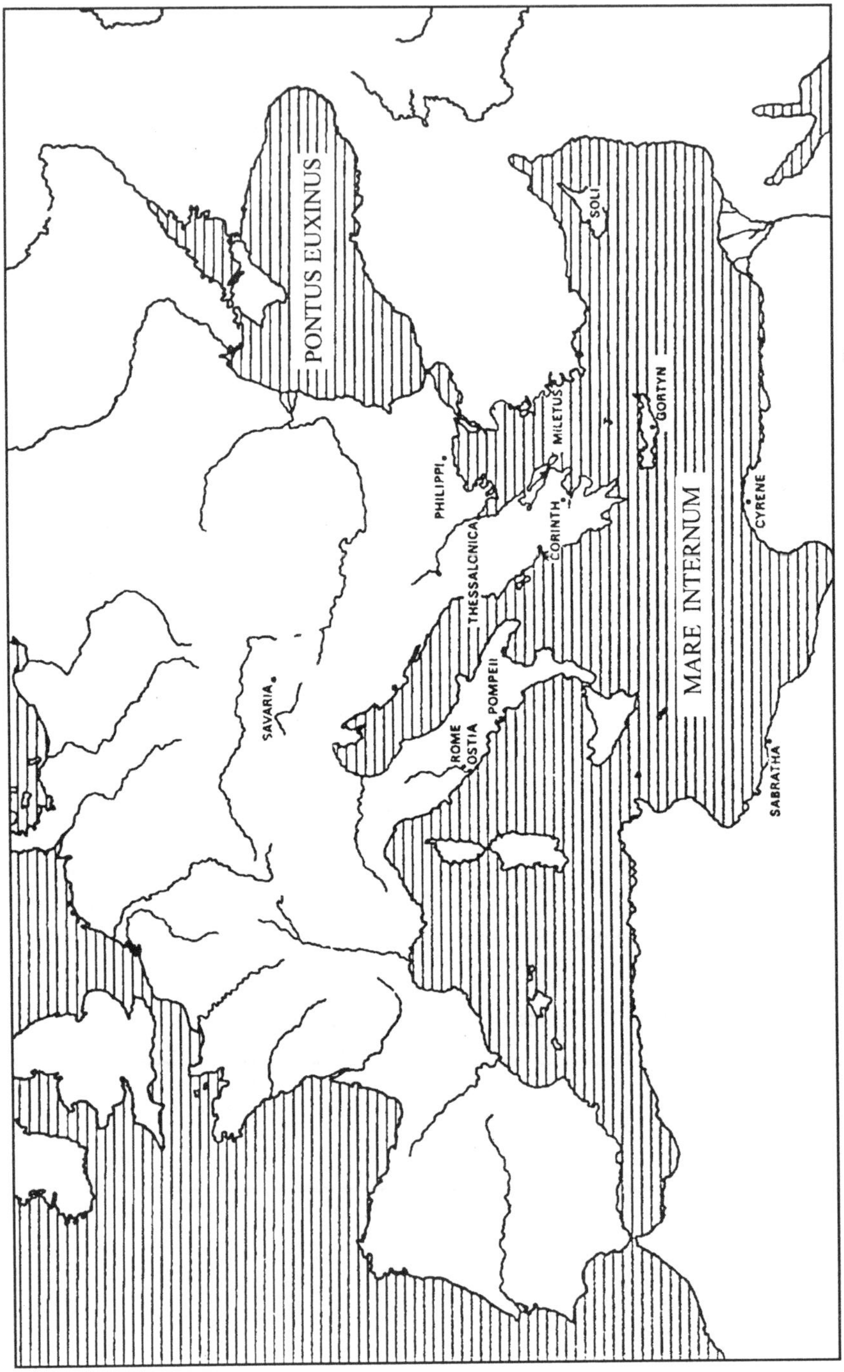

Abb. 1. Die von R. Wild mit Sicherheit indentifizierten Isis-Sarapis Tempel (außer Ägypten)

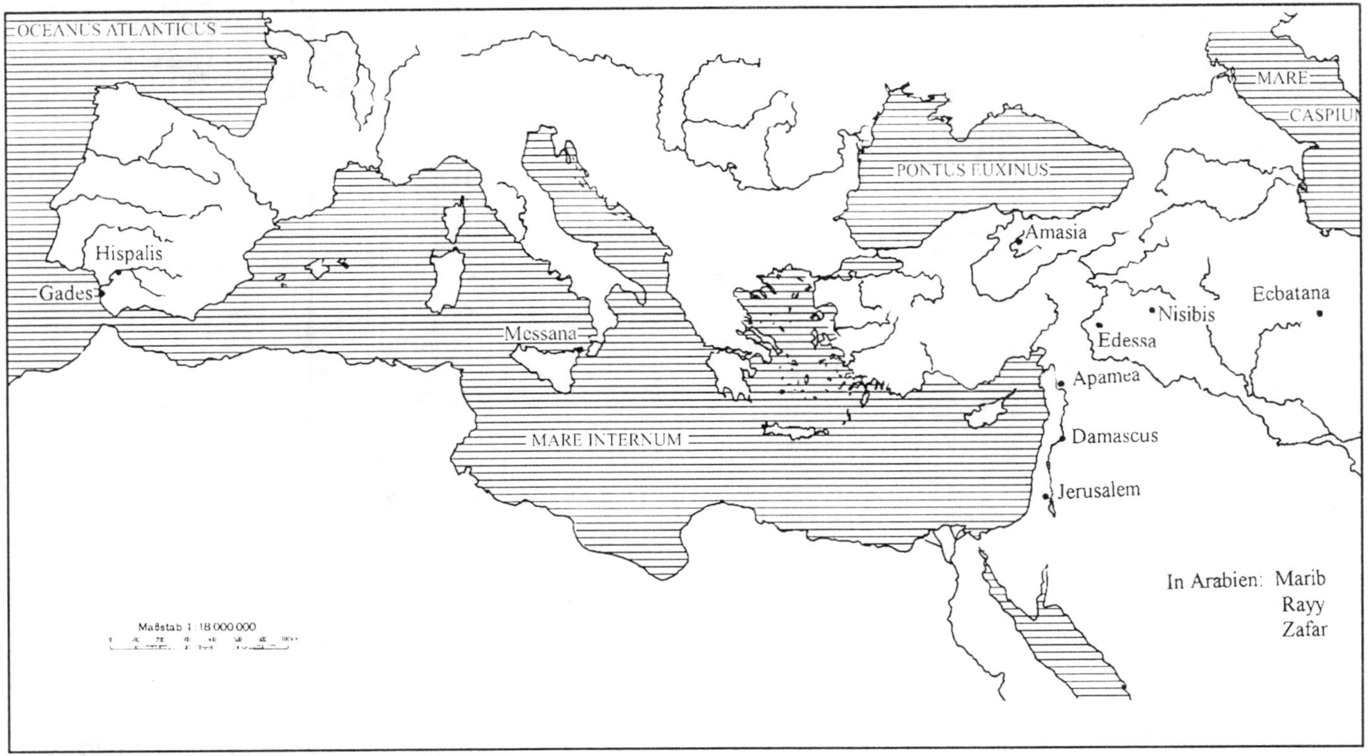

Abb. 2. Großstädte des Mediterraneums in der griechisch-römischen Zeit, wo keine isis-Sarapis Tempel gefunden wurden.

ganz sicher; wenn wir davon die fünf zweifelhaften Tempel abziehen, bleiben nur noch 31 übrig. Abgesehen von den, im letz-ten Halbjahrtausend des Imperiums außerhalb von Ägypten gebau-ten, nur eventuell oder wahrscheinlich der Isis oder Sarapis geweih-ten Tempeln beträgt die Zahl der mit Sicherheit Isis oder Sarapis gewidmeten Tempel insgesamt 13: Korinthos, Kyrene, Gortyn, Miletos, Ostia, Philippi, Pompei, in Rom der auf dem Campus Martius und der der Sta Sabina, Sabratha, Soli und Thessalonika. Ausgesprochen nur Isis geweiht sind davon nur vier, der eine davon ist der, für mich viele Fragen unbeantwortet lassende Tempel in Savaria. Die niedrige Zahl und die geographische Verbreitung sind schon vielsagend und erwecken zugleich Zweifel hinsichtlich des nördlich der Alpen zu findenden und auch mit Ägypten verbundenen Kultes der Göttin Isis und des Sarapis. Das zwei Jahre vor dem Fund in Szombathely entdeckte, damals oft zitierte Iseum in Frauenberg, das die nächste Analogie zu sein schien, wird von Wild nur noch für wahrscheinlich gehalten, seine Beschreibung erscheint eher skeptisch: die Identifikation geschieht auf Grund der Konjektur eines fragmentarischen und schlecht lesbaren ISI zu ISI[DI] auf einem fragmentarischen, eine sitzende Frau darstellenden Relief und eines Wasserspeichers.[29]

In Savaria blieb von der Inschrift im Wesentlichen nur das SAC erhalten, das Tihamér Szentléleky zu [ISIDI AUG](ustae) SAC(rum) ergänzt;[30] vorstellbar ist aber auch [ISIDI] SAC(rum).[31] Ein Unikat stellt die Marmorplatte dar, auf der eine Frauengestalt auf dem Hund Sothis reitet und ein Sistrum hält; bekannt ist jedoch auch die Münze Faustinas II. (???-161), welche die auf einem Hund reitende Isis darstellt.[32] Wir glauben sogar auf Grund

[29] ANRW II. 17/4, 1984, 1778-1779.

[30] SZENTLÉLEKY 1960, 18 wird die Ergänzung als "vermutlich" bezeichnet (so in der deutschen Fassung auch, S. 32.); hnlich in der Ausgabe von SZENTLÉLEKY 1965, 18, deutsch 45.

[31] SIRIS 663; TAKÁCS 1995, 174.

[32] ARSLAN 1997, IV 48; eine ähnliche Münze ließ auch Jovian (363-364), der christliche Nachfolger Julianus Apostatas prägen, der gar nicht mehr das Fest Navigium Isidis am 5. März 364 erlebt, ARSLAN 1997 IV. 122. Bekannt

der Münze Vespasians,[33] daß auch auf dem Giebel des, auf dem Campus Martius stehenden Tempel eine solche, auf einem Hund reitende Isis dargestellt war, welcher jedoch in die entgegengesetzte Richtung schaute).

Problematischer ist der Anubis-Stein, den István Schoenvisner im Jahre 1791 folgendermaßen beschrieb: "Lapis quadratus, in quo Anubis, seu Cynocephalus repraesentatur inter duo alia Numina medius, – Isidemne, Serapidem? an Hygiaeam et Aesculapium dicam? incertus haereo."[34] In den vergangenen zweihundert Jahren wurde diese, aus der "Prähistorie" der Ägyptologie stammende, die Ikonographie des Sarapis noch nicht kennende Definition nicht weiter untersucht[35] und sogar die Sarapis-Linie als sicher angenommen. In Anlehnung an die Idee Dobrovits'[36] hielt Wessetzky die mittlere Figur, die in ihrer Rechten einen caducaeus und inihrer Linken einen Palmzweig hält nicht für Anubis, sondern für eine als Anubis maskierte Gestalt, einen Priester. Wenn es sich um einen Priester handelt, dann bilden auch die zwei Gestalten den Teil einer Prozession, der Prozession Navigium Isidis.[37] In diesem Falle müssen wir freilich die sich ergebende innere Diskrepanz berücksichtigen, bisher wurden noch nirgendwo Darstellungen maskierter Isis- bzw. Sarapis-Gestalten in einer Prozession gefunden, dies jedoch entzieht einer traditionellen Interpretation den Boden. Für mich ist auch fraglich, welchen Sitz im Leben das Fest der Seefahrt in Savaria gehabt haben sollte. Es gibt aber auch noch andere Fragen. Über den stehenden Sarapis existiert schon eine

ist noch ein ähnliches Bronzefigürchen (3,9 cm, Vatikan) ohne Herkunft und Zeitangabe, ARSLAN 1997 IX. 3., S. 623.

33 MALAISE 1972, No. 391 und Tf. XXX. 29 ibid. 1778-1779. 34 Antiquitatum et Historiae Sabariensis ab Origine usque ad Presens Tempus Libri Novem, zitiert von WESSETZKY 1961, 30 und WESSETZKY 1989, 22.

35 In neuerer Zeit erwähnt in: KATER-SIBBES 1973, No. 875; GRENIER 1977, 150 No. 233, XXIII, TRAN TAM TINH 1983, 137-138, Fig. 78.

36 DOBROVITS 1943, 49.

37 WESSETZKY 1959, 265 ff., WESSETZKY 1959b, 22.

Monographie,[38] aus der wir wissen, daß er, der kanonischen Darstellung des Gottes entsprechend mit seiner Linken das Zepter hält (oder sich einen langen Stock stützt) und seine Rechte auf dem dreiköpfigen Cerberus ruhen läßt – mit seiner Rechten stützt er sich also nicht einen Stock wie auf dem Stein in Savaria. Religiöse Traditionen sind außergewöhnlich konservativ – so wie die Ikonographie eines christlichen Altarbildes nur innerhalb der festgelegten Grenzen vorstellbar oder für uns das "er sitzt zur Rechten Gottes, des allmächtigen Vaters" im Credo nicht auswechselbar ist, so wird sich auch die Sarapis-Tradition an ihre üblichen Vorbilder gehalten haben.[39] Problematisch ist nicht nur die Männergestalt auf der rechten Seite, sondern auch die Frauengestalt auf der linken Seite. Tihamér Szentléleky[40] beschreibt sie als Isis, das Haupt mit einem modius bedeckt, mit der sich windenden Schlange in der Hand. Das wirkliche Problem besteht nicht darin, daß mir eine Darstellung der Göttin mit modius nicht bekannt ist - dies ergibt sich ohnehin nicht zwingend aus dem Fragment -, sondern in der funktionalen Deutung der Schlange in der vorliegenden Anordnung und andererseits in der gesamten Komposition. Äußerst selten vorzufinden ist eine solche Darstellung, welche die Trias zusammen darstellt (vor allem nicht auf einem Tonstempel oder auf einer Münze, sondern in einer Großplastik[41]), noch seltener ist die Placierung des Kindgottes in der Mitte. Eine Darstellung jedoch, auf der zwischen beiden Anubis steht, kenne ich nicht. Ob sich die

38 TRAN TAM TINH 1983.

39 Bekannt sind zwei Ausnahmen innerhalb eines corpus von mehreren Hundert: ein Altar im Museum von Liverpool (TRAN TAM TINH 1983, Tf. XXXV-XXXVII.) und Münzen (TRAN TAM TINH 1983, II. 9-27) aus der Zeit des Marc Aurel, Commodus, Elagabal, Severus Alexander, Maximinus I., Gordianus III., Philippus I., Decius, Aurelian und Diocletian.

40 SZENTLÉLEKY 1960, 20. (deutsch S. 32.) Hier und in seiner ähnlichen Studie von 1965, 20. (deutsch S. 45.) bezeichnet er die rechte Gestalt ohne Begründung als Osiris.

41 Auf einem Relief in Kairo steht zwischen Isis linkerhand und Sarapis rechterhand Horus, TRAN TAM TINH 1983, 89, Nr. IA2.

Schlange von der rechten Gestalt her oder auf sie zu windet, kann auf Grund der Darstellung schwer entschieden werden. Dies kann nur in Abhängigkeit von einer bereits geklärten Ikonographie und in Anlehnung an eine Deutung verantwortlich ausgesagt werden.[42] Mit den Worten Schoenvisners: "incertus haereo", eine nicht zufriedenstellend identifizierte Darstellung kann nicht einer anderen unsicheren Identifikation als Grundlage dienen. Man müßte vielleicht ein Neudurchdenken der Interpretation versuchen oder die bisherigen Behauptungen durch eine kohärente Argumentation stützen.

Meiner Ansicht nach müßte am ehesten der Beweis dafür gefunden werden, ob die nicht ganz geklärten Marmortafeln, wenngleich sie sich auch am Fries des Tempels befinden, wirklich nur dort placiert werden konnten, wo sie heute eingefügt sind beziehungsweise ob die vorliegenden Victoria-, Fortuna-Abundantia-, Mars-Harpokrates- und Hercules-Darstellungen (sowie andere, damals sicher vorhandene, heute nicht mehr existierende Reliefs) nicht eine andere Interpretation zulassen, derzufolge die auf Sothis reitenden Isis und Kynokephalos nur συνναοι θεοι gewesen wären. Das heißt: Waren die Herren des Tempels eindeutig ägyptische Götter (bzw. nur eine Göttin) oder aber waren neben den römischen Göttern noch weitere, romanisierte orientalische Götter anwesend? Von entscheidender Bedeutung wäre auch, wenn wir die zentrale Darstellung des Frieses über dem Tor des Tempels kennen würden.

Den 1831 in Egyed (Komitat Győr-Sopron) gefundenen unvergleichbar schönen Krug, gemeinsam mit der gleichzeitig zutagegetretenen ägyptisierenden, jedoch einen anderen Stil tragenden patera, brachte Vilmos Wessetzky mit dem Kultzentrum von Savaria in Zusammenhang.[43] Aus diesem Grund ist für uns an dieser Stelle nicht die herausragende ästhetische Qualität wichtig, sondern die kultische Funktion. A. Hekler lenkt die Aufmerksamkeit

42 Wohl kann auf der cista mystica ein Schlange sein. Die Darstellung der Isis mit einer Schlange in der Hand ist noch von Stelen bekannt. Schwer zu entscheiden ist, ob es sich auf dem, eine Isis-Prozession darstellenden Relief im Vatikan um eine Schlange oder einen schlangenförmigen Armreif handelt.

43 WESSETZKY 1959, 265ff, WESSETZKY 1959b, 27 und WESSETZKY 1961, 44.

auf das Relief mit der Isis-Prozession im Vatikan und die sich darauf befindliche, ähnliche Hydria, Wessetzky hingegen erwähnt, daß die patera das aus dem Krug gegossene Wasser aufgefangen haben mag[44] – dies sind zwei verschiedene Konzeptionen. Hekler beschreibt,[45] daß die Hydria "in drei Teilen gefunden (wurde). Fuss und Mündungsstück sind aus Bronze, der Gefässkörper aus Kupfer. An der Zusammengehörigkeit dieser Teile zu zweifeln ist kein Grund vorhanden." Demzufolge hat sie ihre heutige Gestalt im 18. Jahrhundert zurückerhalten.Gleichzeitig akzeptiert er aber nicht, daß der mit ihr gleichzeitig aufgefundene, grob ausgeführte Henkel zur Hydria gehört, selbst wenn er passen würde, würde die unterschiedliche Qualität gegen eine Zusammengehörigkeit sprechen. Der Krug hat bis zum heutigen Tag keinen Henkel. Wir sehen also, daß Hekler als Richtschnur den qualitativen Unterschied ansetzt. Darüber, daß der, an die patera angefügte Bronzegriff zu dem Kupferteller sekundär ist, sind sich alle einig. Dies sollte meiner Ansicht nach auch in Bezug auf den Krug durchdacht werden. Hekler leitet nämlich die kultische Funktion des Kruges von der Oinokhoe-Form ab.

Wenn wir nun aber den aus fremdem Material hergestellten Schnabel und Fuß entfernen, erhalten wir eine ganz andere Form. Es ist merkwürdig, daß bei dem mehrfach untersuchten Krug nie aufgefallen ist, daß auch die Ornamentik der aus unterschiedlichem Stoff hergestellten Teile unterschiedlich ist – eine ägyptische Darstellung gibt es nur auf dem Bauch des Gefäßes, die Bronzeteile werden von einer pflanzlichen Ornamentik geschmückt. Der Ausguß und der vormals dazugehörige Henkel, dessen grobe Schweißnähte bis heute sichtbar sind, existierten ursprünglich nicht und konnten auch gar nicht für den Eimer konzipiert gewesen sein, denn die, auf dem Mantel des Mittelteils befindlichen acht göttlichen Gestalten sind mit mathematischer Genauigkeit genau in einem Winkel von 45 Grad angeordnet, für einen Henkel hat der Künstler keinen Platz eingeplant. Er konnte

[44] WESSETZKY 1989, 32.
[45] HEKLER 1909, 36.

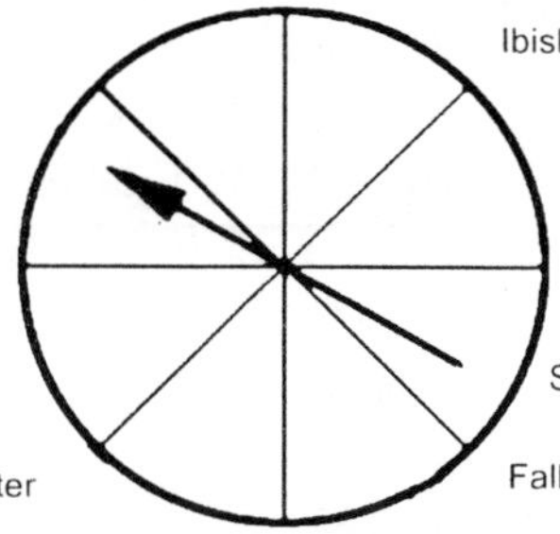

Abb. 3. Das Verhältniß des Schnabels zum Götterfries

auch nicht wissen, daß später ein ungeschickter Mensch, mit Müh und Not die Figuren nicht verletzend, zwischen den linken Ellenbogen des falkenköpfigen Gottes und der Flügelspitze der ihm den Rücken zuwendenden Göttin etwas anschweißen wird, wodurch der Schnabel des Oinokhoe im Vergleich zu den Darstellungen in eine asymmetrische Lage kommt (Abb. 3 und Tf. 22). Dies widerspricht in jedem Fall der Qualität des ersten Meisters. Nach diesen Erwägungen kommt mir der Krug ein wenig ungeschickter vor und ich kann darin nicht mehr die unvergleichliche Harmonie und die vollkommenen Maße sehen wie einst. Zum augenblicklichen Zeitpunkt kann ich nur so viel sagen, daß das Gefäß ursprünglich kein Oinokhoe war, ich möchte jedoch die Schlußfolgerungen nicht übereilen. Es ist möglich, daß dies auch etwas über die Funktion aussagen wird. Möglicherweise verändert sich auch die Datierung: Der kleine kupferne Stamnos und die sekundär angefertigte Oinokhoe erhalten eine eigene Datierung. (Die einstige Fundgruppe läßt sich nicht mehr vollständig rekonstruieren, ich kann aber nur hoffen, daß der heute bekannte Kunstgegenstand nicht erst im 18. Jahrhundert angefertigt wurde.)

Die ägyptische Darstellung muß noch gründlich untersucht werden, da bisher nur Stellungnahmen zu einzelner Fragen geäußert wurden. Solange wir nur sagen, daß die Göttinnen nur möglicherweise Isis darstellen und der ibisköpfige ist Thot (auf die Zweiheit gibt es keine beruhigende Antwort) und den Neujahrsfrosch für den dominanten Darsteller der Szene halten, solange halte ich für bedrückend, daß wir den Sinn des gesamten Frieses nicht im Kanon der ägyptischen Darstellungen gefunden haben. Der kupferne Stamnos ist ganz sicher eine ägyptische Arbeit, es ist jedoch unwahrscheinlich, daß es dies auch in seiner heutigen, ergänzten Form ist.

Freilich kann in einem fraglichen Tempel einer sekundären Isis-Religion der Gebrauch eines sekundären Stücks nicht ausgeschlossen werden, bevor wir jedoch Hypothesen aufstellen, muß noch viel Forschungsarbeit geleistet werden.

Peter Hubai
Ägyptische Sammlung
Museum der Bildenden Künste
Budapest

ISIS IN ROM

Als Ägypten den Tiefpunkt seiner Geschichte erreicht hatte und unter den Provinzen des Römischen Reiches als privater Besitz des Kaisers in die unterworfenste Position geraten war, sind die Götter des niedergetretenen Landes auf paradoxe Weise in die Provinzen des Imperiums und auch selbst in Rom siegreich eingezogen. Weder die politischen Bestrebungen des Augustus für die Wiederherstellung der altrömischen Sitten und Religion noch die von rational-philosophischen Tendenzen entkräfteten griechischen Kulte konnten die Verbreitung der Götter aus dem Osten aufhalten.

Siegfried Morenz hat darauf hingewiesen, dass, obwohl die Verehrung der Götter Ägyptens in Laufe der ersten Jahrhunderte der Kaiserzeit von einer ethnischen Religion zur echten universalen Religion geworden war, doch soll man nicht von einer echten Isis-Mission sprechen.[1] Nicht Isis wollte über die griechisch-römische Welt regieren, vielmehr hat diese Welt Sie gebraucht.

Die ägyptischen Götter sind von ihrer Heimat zuerst während der Zeit der Republik und in der frühen Kaiserzeit aufgebrochen und mit gräzisierten Namen ein hellenisiertes Aussehen erhalten. Eine gegenseitige ägyptisierende Wende auf diesem Gebiet ist, – nicht zufällig – unter der Herrschaft Domitians eingetreten, denn er war es, der das Iseum Campense in prächtiger Gestalt wieder aufbauen liess.[2] In Ägypten

1 MORENZ 1975, 521-526.

2 MALAISE 1972a, 414-417

sind die lokalen Kulte in dieser Zeit auf spektakuläre Weise wieder aufgelebt. Dies ist auch in der Münzprägung durch die Nomosmünzen belegt.[3] Während früher nur die Bilder der grossen Götter auf Münzen erschienen, wurden von nun an Münzen in Umlauf gebracht die lokale ägyptische Gottheiten darstellten.

Domitian hat offenbar ägyptische Gelehrten mit dem Verfassen der hieroglyphischen Inschrift seines riesigen Obelisken beauftragt,[4] der jetzt auf der Piazza Navona steht und als eine der berühmten Sehenswürdigkeiten von Rom gilt. Das Ansehen der Hieroglyphen in Italien ist auch durch zwei andere mit hieroglyphischen Inschriften versehene Obelisken bezeugt, die sich in Benevent befinden. Diese Denkmäler zeigen auch die veränderte Funktion der Obelisken. In der Zeit des Augustus haben die zwei nach Rom verschleppten Obelisken den Sieg über Ägypten propagiert,[5] jetzt sind sie dagegen Anzeichen des Ruhms der ägyptischen Götter und des Imperators, der den ägyptischen Göttern huldigt, sowie Symbole der Wiederbelebung der uralten Kultur und des Interesses an den Denkmälern im Niltal. Vielleicht noch unter Domitian oder während der Zeit des Hadrians war Hermapion tätig, der die durch Ammianus Marcellinus tradierte Inschrift von einem der römischen Obelisken ins Griechisch übersetzt hat.[6]

Vom Ende des ersten Jahrhunderts an ist nicht nur die Lehre, der religiöse Inhalt für die Gläubigen wichtig, sondern auch das heilige Land selbst. Man wollte ägyptische Objekte haben, um den Isis-Kultstätten ägyptisches Aussehen zu verleihen. Echte ägyptische Statuen sollten die couleur locale in die Tempel zaubern,[7] und die Priester sollten die liturgischen

[3] MILNE 1924, 33-34

[4] IVERSEN 1968, 203-207, Tf. 20-21.

[5] KÁKOSY 1967, 314.

[6] Ammianus Marc. XVII.4.18

[7] LEMBKE 1994, 202-223.

Gebete, – wie wir es bei Apuleius lesen, – von hieratischen Papyri in der heiligen Sprache rezitieren. Wenn nicht genug ägyptische Werke zur Verfügung standen, sollten lokale Künstler Imitationen besorgen, wie es die Villa Hadriana zeigt. Die letzte hieroglyphische Inschrift, die noch die souveräne Behandlung der alten Schrift und die eingehende Kenntnis der klassischen ägyptischen Sprache sowie die Fähigkeit einen längeren religiösen Text zu schaffen bekundet, ist die Inschrift des Antinous-Obelisken aus der Zeit des Hadrians, der jetzt auf dem Monte Pincio steht.[8]

Die pharaonische Renaissance, die Sehnsucht nach der alten Weisheit, konnte sich in erster Linie im italischen geistigen Klima durchsetzen, blieb aber nicht ohne Wirkung selbst in ferneren Gebieten, wie Pannonien. Die Neujahssymbolik der Kanne von Egyed bewahrt noch die ägyptische Geistigkeit und der Fries[9] mit den ägyptischen Diademen bezeugt noch die gründlichen Kenntnisse der alten Ikonographie.[10] Der bei Balatonszabadi gefundene Altar konnte – trotz aller Primitivität – nur von einem Steinmetz geschaffen werden, der eine gewisse Vertrautheit mit dem hieroglyphischen Schriftsystem besass. Er weiss z. B. wie man den Namen des Osiris schreibt.[11]

Die Anziehungskraft der Isis-Religion wurde durch die Mysterien gesteigert, die auf die letzten Fragen des Seins eine Antwort, ein geheimes Wissen versprachen, das das ganze Leben des Eingeweihten verändern konnte. Die Mystereien hatten mehrere Stufen. Der Einweihung ist in Metamorphoses des Apuleius eine spontane Erleuchtung vorausgegangen, nicht im verschlossenen Sanktuar des Tempels, sondern in der freien Natur. Lucius empfindet, noch in Eselsgestalt, die Allmächtigkeit der Göttin an der Meeresküste, im strahlenden Licht des Mondes, als er der göttlichen Vision noch nicht teil-

[8] GRIMM-KESSLER-MEYER 1994; IVERSEN 1968, 161-173.

[9] KÁKOSY 1983, 59, Tf. 2.2/3.

[10] WESSETZKY 1961, 44, Abb. 10-12.

[11] WESSETZKY 1961, 47-48, Abb. 14-16.

haftig war und die Offenbarung aus dem Mund der Isis noch nicht erhalten hatte.

Die Anweisungen der Göttin folgend, befreit er sich vom verhassten Äusseren des Esels, das heisst des Seth. Dann wird er stufenweise des beglückenden Erlebnisses der Einweihung würdig. Da lernt er das Universum, das Jenseitsreich kennen. Er sieht die Sonne in der Unterwelt strahlen,[12] und später am Morgen nach der Einweihung , im Zuge der solaren Verklärung, wird er mit dem von Totenbuch Kapitel 19 bekannten Kranz der Verklärung auf dem Kopf vor den Isis-Gläubigen erscheinen.[13]

Die ältere Meinung der Forschung, dass die Einweihung und Mysterienkult nur ein späteres, römerzeitliches Phänomen in der ägyptischen Religion sei, und dass wir darin nur die Übernahme der eleusinischen Mysterien sehen sollten, muss revidiert werden. Die Enthüllung heimlichen Wissens, die Schau heiliger Objekte, wie der Tag- und Nachtbarke der Sonne oder das Zeigen der im Sanktuar aufbewahrtenn Kultstatuen für einige Privilegierten sind bereits gut bekannt von den älteren ägyptischen Tempeln.[14]

Neben der rituellen Einweihung gibt es wenige Worte für die theurgischen Liturgien als spiritueller Einweihung. Der neoplatonische Philosoph, Porphyrios beschreibt im Lebenslauf seines Meisters Plotinos[15] einen okkulten Ritus, dessen Haupt-darsteller ein ägyptischer Priester war, der im Iseum in Rom den Geist des Plotinos heraufschwörte und herausfand, dass der Körper des grossen Philosophen der Wohnsitz eines Gottes war.

Wir wissen von Plutarch, dass der Philosoph das Erlangen des esoterischen Wissens durch geistliche Erleuchtung zu erhalten erhoffte. Es ist der heimlichste Teil der Philosophie

12 Apul. Met. XI.23

13 Apul. Met. XI.24.

14 KÁKOSY 1999, 160-161.

15 Porph. Vita Plot. 10.

(Lesemysterium), dessen Besitz dadurch erlangt wird, dass die Erkenntnis, das echte Wissen, die Seele wie ein Blitz durchzuckt.[16]

Ohne Beweismaterial können wir nicht behaupten, nur vermuten, dass auch die Isis-Gläubigen in Pannonien allein durch die Bewunderung der Pracht des Kultes nicht befriedigt wurden, sondern tiefer eindringend in die unerschöpfliche Vielfalt des Wesens der "tausendnamigen" Göttin die vollständige Wahrheit kennenlernen wollten.

László Kákosy
Universität Eötvös Lóránd
Budapest

16 Plut. De iside 77.

A New Concept for the Reconstruction of the Iseum of Szombathely

Foundations as revealed by the 1955-1962 season

Hajnóczi's reconstruction

The 1955-1962 seasons of excavations unearthed the temple's walls foundations and the temenos located around the temple. This construction measures 50x70 ms[1] and can be divided in two different areas. For instance, along the Roman roadway, there is a narrow passage confined what could have been a portico. Behind this building there is a large hall measuring approximately about 15 ms in width and which runs along the entire length of the temenos. The hall was divided by granite columns which bases can still be seen in situ. However, it is unclear as to whether this space was simply a vestibule for the temenos, enclosed by cross-walls, or if it was a large open space that ran all the way to the streets and to the sacred place. The temple is located in a 30x50 ms courtyard.

The buildings that flank the two long sides of the courtyard have been identified as some sort of shops. At the temple's entrance there is a sacrificial altar. Behind the shrine, the temenos southwestern wall runs for about 12 ms. From here, the remains of the road's edge can be seen for about 28 ms. Similarly, along the length of the southeastern side Roman road's edge runs about 12 ms from the outside tabernacle's wall. There

[1] HAJNÓCZI - SZENTLÉLEKY 1964, 129-135. The measurements of the temple are given in the article p. 132.

are visible remains on the northwestern side of the road and the temenos, which is covered by the fences of modern homes. Meanwhile, on the southeastern and southwestern sides, the distance from the wall to the road is so significant that some sort of constructions may have been existed there. The shrine's foundation walls have remained almost intact.

The missing pieces of the northwestern corner's wall are the result of World War II activity. Excavations have successfully revealed the walls surrounding the *cella,* the podium temple's porch (block remains on the northeastern corner of the shrine's foyer) and five cross walls perpendicular to the temple's length ways axis. Available data suggests that what Hajnóczi has determined to be the *cella* was constructed with very sturdy and thick walls.

In the shrines's foyer (the *pronaos* area, according to some current reconstruction), the walls that have been excavated are of an inferior quality and in much more worse condition than the walls in the *cella*. The highest place located on the northeastern side of this complex of walls, presents remains of what appear to be floor tiles. Direct view of this feature is actually no possible, due to the reconstruction of the *pronaos*.

During the last part of the 1950's decade, approximately 30%[2] of the site was successfully cleared and now, four decades later, a new approach is necessary to determine the sites's different periods.

The problems of mass reconstruction

The main facade's theoretical reconstruction is based on fragments obtained from the frieze. Hajnóczi determined the placement of the front columns based on fragments found also on the frieze. He set the outermost row of columns[3] at a distance of

2 Tihamér Szentléleky mentioned the data during the symposium.

3 According to the commentary written on the fourth theoretical plan published in November 30, 1960: "Data concerning the width: the portico's two extreme rows of columns, based on the width of the relief are at 2.24 ms. The

about 2.24 ms[4], the center row at a distance of about 2.72 ms and the axial distance at about 3.24 ms. Unfortunately it is not possible to examine the marble reliefs at a close distance, because they have been placed at their original heights on a reinforced concrete structure. Endre Tóth, István Tóth, and most recently Zsolt Mráv in the course of their research logically question the plausibility of certain placements. From the placement of the columns arises the question of whether the reconstruction is unconventional, as the *pronaos* and the *cella*'s width are not identical[5] and have a 50 cm difference. Moreover, the "bun-like" shape of the roof does not conform at all with Roman architectural style. As it is unthinkable that the gabled roof, enclosing the tympanum and built over the width of the *pronaos*, would connect to a wider gabled roof. It is clear and logical that Gyula Hajnóczi asserted the existence of an attic-like roof after he determined the width of the *pronaos*. The architectural problem of the attic could have been settled with, a now lost, small piece of the cornice.

In the course of Zsolt Mráv's research, it has been shown that the area inscriptions would have required a wider area. Further consideration and debate are needed over the placement of the frieze on the surface. Furthermore, the reconstruction of the frieze leads to another key question: whether the features have been correctly placed in context with each other. According to Tihamér Szentléleky, the fragments of the

middle one, determined by the symmetry of the ornaments is at a 3.24 ms distance. This data does not still give the temple's width. Probably the portico was narrower than the temple. The thickness of the walls cannot be less than the recommended 30 cm."

4 Correctly it is not the distance between the rows of columns that is 2.24 ms but rather the distance of the axis of the extreme columns. The distance here is according to this plan 1.72 ms.

5 In the second part of the publication Szentléleky theorises that "the width of the entry hall was smaller than that of the cella, but, thus, it seemed to leap out of the mass of the temple..." HAJNÓCZY – SZENTLÉLEKY 1964, 134.

frieze fit together. István Tóth says that the illustrations do not rightly fit next to each other.[6] On the other hand, Ottó Sosztarits claims that the difference between the frames of the structures justifies the need for altering the reconstruction.[7] In

Fig. 1. The plan of the shrine, after the excavations

addition, it is not possible to study the relief work in its present position and it is questionable if it will ever be due to its cemented-situation. They will be damaged during their extraction from the concrete and the extent of this damage cannot be predicted. Thus, we will depend on the results of a stylistic assessment during the preparation for a new reconstruction. A similar problem arises with regard to the relationship between the stairway and the *pronaos*. I maintain that the stairs' position in relation to the wall'facade is untenable. Nevertheless, looking at the remains of the wall, it is not possible to change the stair's location (Fig. 1).

The concept of the interior space constructed with barrel vaulting also raises doubts. The fourth plan and its commentary as shown in the first annotation are problematic due to the basic principles of stone wall construction. A stone wall that measures 30 cm

6 TÓTH 1996-97, 331

7 Personal communication by I. Tóth, partially published in: TÓTH 1996-1997, reference to p. 332, also Sosztarits' own statement.

in thickness is simply not possible, since at least it should be 40 cm thick. Studies of Roman walls' remains suggest that walls that measure 35-40 cm in thickness are of a secondary importance and not weight bearing. Barrel vaulting correlated in width with the cella would have required for walls of about 80 and 100 cm in thickness.

The shrine of Savaria interpreted by function and analogy

The necessary correlation between function and architectural space and the needs for the Isis' cult liturgy call for certain principles that justify a complete revaluation of Hajnóczi's reconstruction. For instance we have not succeeded in finding the scribe's room mentioned in the descriptions of the cult. Neither in Plutarchus "Isis and Osiris", nor in Apuleius "The eleventh and closing book" have useful description of it. Nevertheless, Apuleius shows that in the procession at least 10 priests had to emerge from the shrine when, on the nineteenth day of the month of Athur, they moved towards the sea. The ceremony meant to drive away drought and clearly, it also took place here in Savaria. At this time perhaps, similar scenes depicted in the frescoes of the Herculaneum walls show that the gods "stood" in the foyer, on the stairs and in the *pronaos*. Meantime at the altar, water was poured from a pitcher into a casket containing sand from which a human figure was kneaded as "a way to find Osiris".

In order for the faithful to witness this ritual it was necessary for the *pronaos* and possibly a set of steps, to be positioned in front of the *cella*. Perhaps the altar was on a different level from the area surrounding it,[8] raised in order to be at the same level as the podium ambulatory, or even higher, neutralizing then, the meaning of the podium. On the other hand, if it

8 "...all the people gave a great shout, and filled with much joy, baring all kind of herbs and garlands of flowers in their way home, kissing and embracing *the steps where the goddess had passed...*". (emphasis by the author) APULEIUS, Metamophoses, p. 247

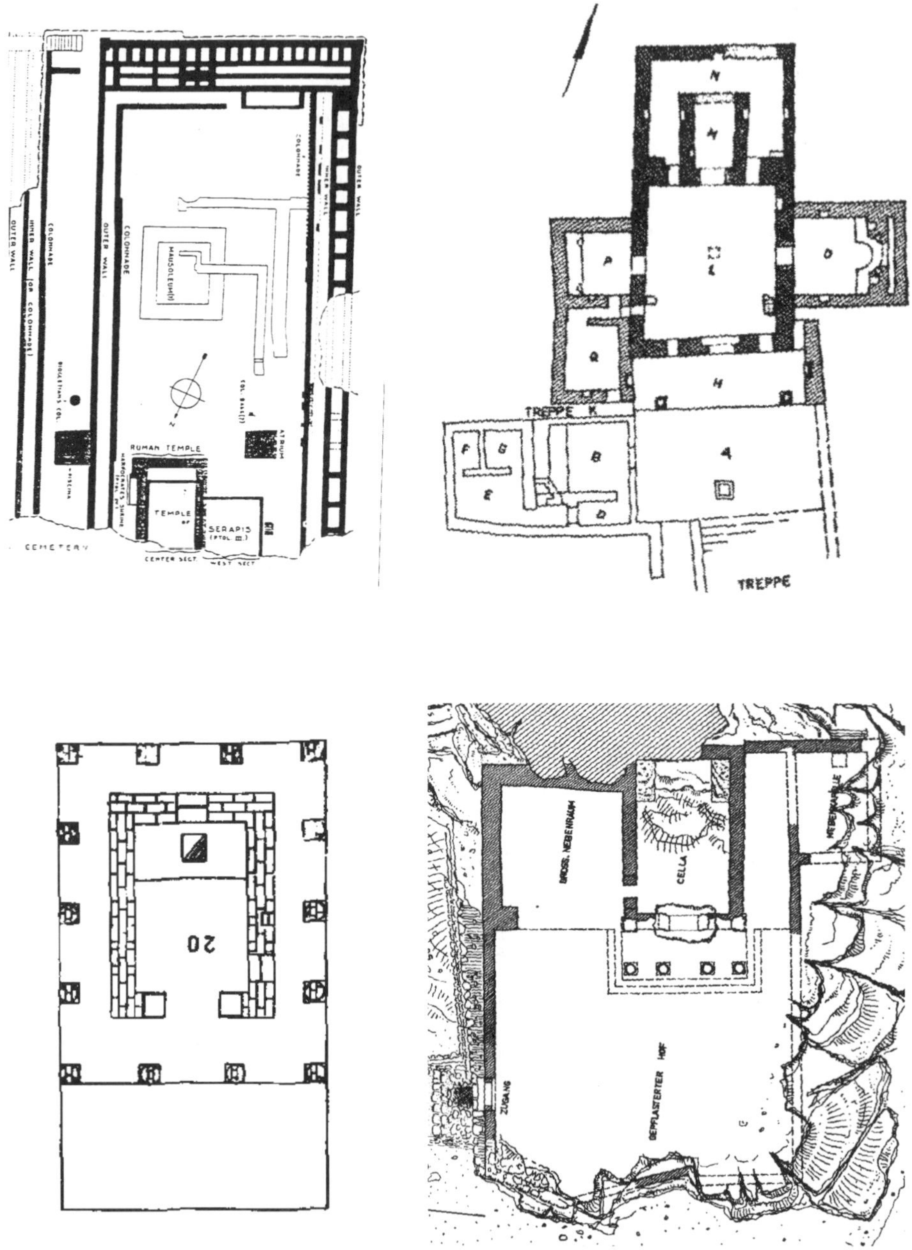

Fig. 2. a. Alexandria, Serapeum b. Mons Claudianus, Serapeum
c. Luxor, Serapeum d. Mons Porphyrites, Serapeum

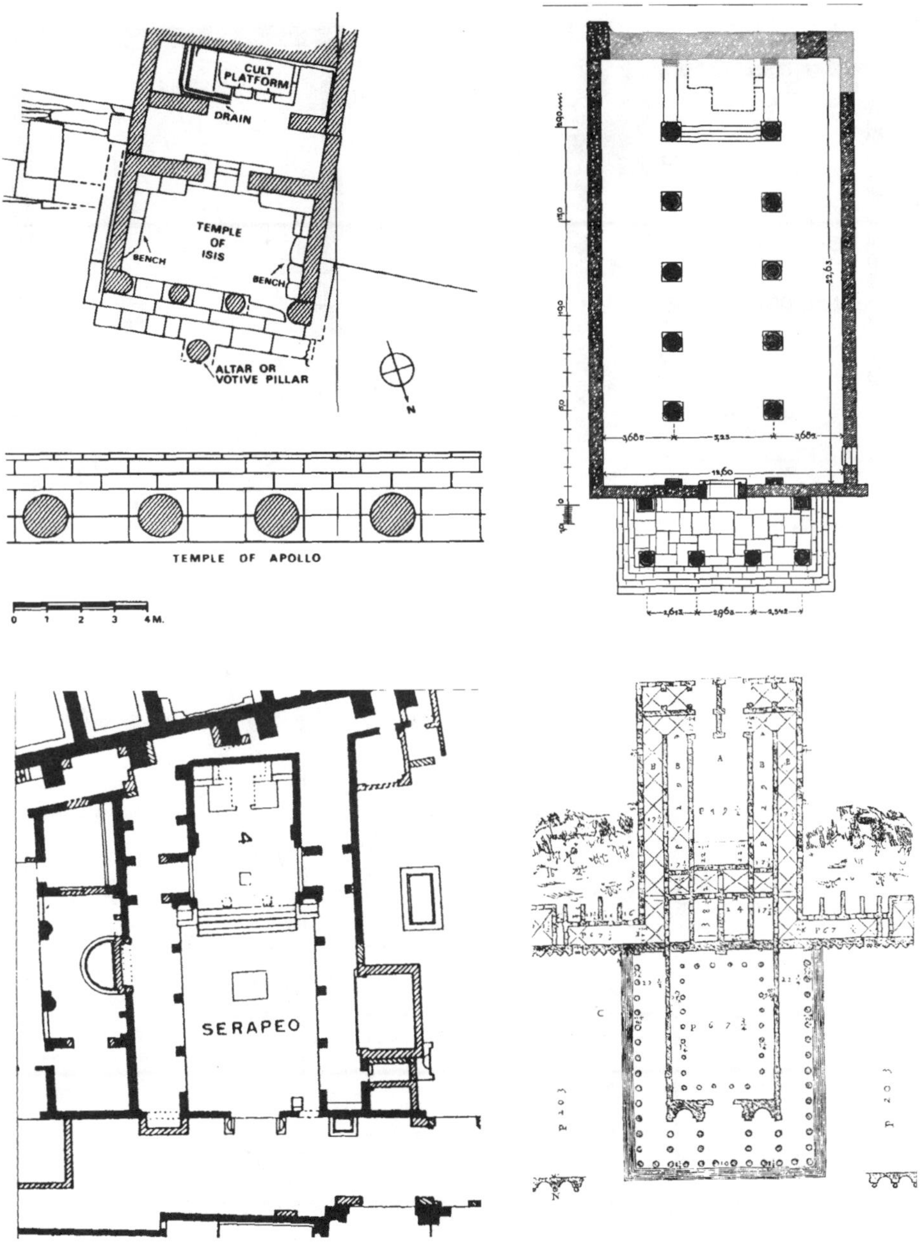

Fig. 3. a. Cyrene, iseum
b. Miletos, Serapeum
c. Roma – Quirinalis, Serapeum
d. Ostia, Serapeum

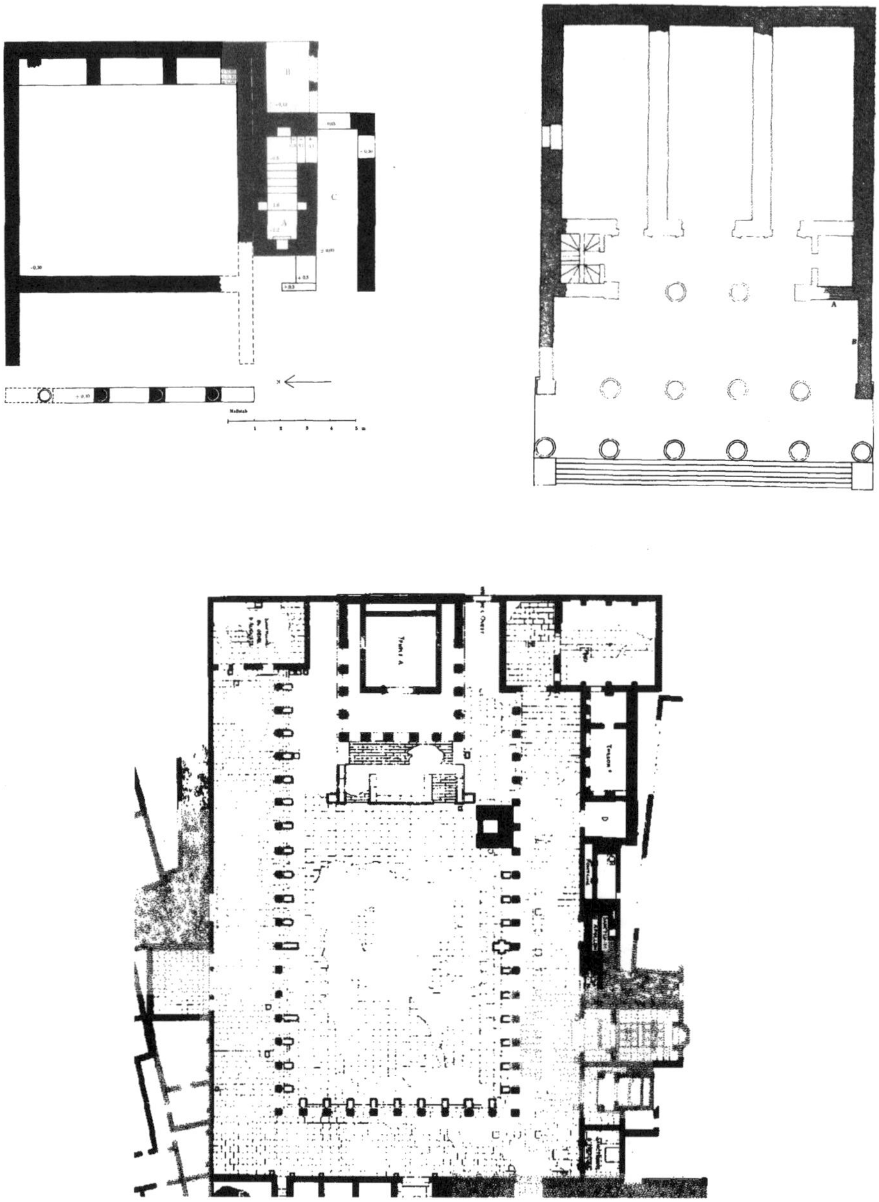

Fig. 4. a. Stratonicea, Serapeum b. Gighthis, Serapeum
c. Gortyn, Serapeum

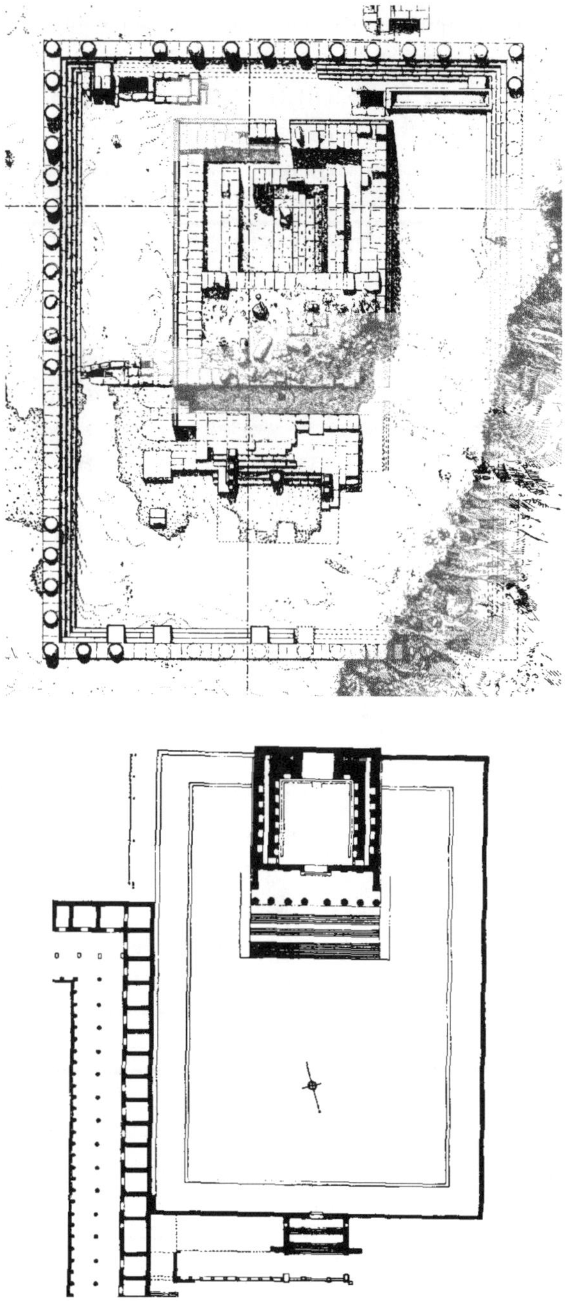

Fig. 5. a. Sabrathi, Serapeum b. Ephesos, Serapeum

did not followed the elevation of the surrounding ambulatory, it could have been in a deeper position. The position of this structures could explain the function of the two retaining walls of the foyer mentioned earlier. It is a fact that the foundation walls of the stairs and the *cella* are made of a completely different quality. It is not clear, with information we actually have, to establish to which type of structure the masonry walls leading to the altar are similar to.

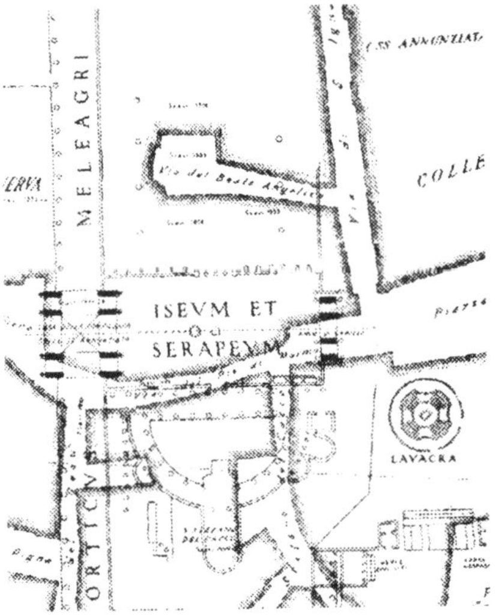

Fig. 6. The site of the Iseum at Serapeum in Rom on the Campus Martius

Plan changes during the expansion of the cult

Figures 2-4 and plates 23/3-26 show the plans of the sixteen excavated buildings and also the wall paintings of the Herculaneum. All the plans, except three of them, are of the Serapeia. Based on this information, a valid canonical plan on the area of the ritual cannot be formulated. The examples are only considered to be representative, because Tim Hegedűs in his paper "The Urban Expansion of the Isis Cult,"[9] asserts that there are a total of 44 ancient cities from Alexandria to Rome in which the cult had a temenos, but knowledge of the plans are limited.

Certainly, there were more shrines in Rome, but unfortunately we do not know much about the main church among them on the Campus Martius. We only know about the facade of this church from a small image depicted on a coin (Fig. 6 and Pl. 23/1-2). Of all the shrines shown, only the one located in Pompeii had a stair that

9 HEGEDUS 1998.

would have been comparable with Hajnóczi's theoretical reconstruction, but even here the stairs were built in front of the mass of buildings and not in between the columns leading towards the *pronaos*.[10] In the cases of the shrines from Ostia, Sabrathu, Gighithis, Leptis Magna, Ephesus, Stratoni-cea and Ras el-Soda, the stairs leading to the temple were the entire width of the building's main facade. This feature is also illustrated in the wall paintings of the Herculaneum. It is not possible to determine the shape of the stairs in the other temples. In some cases, a larger *vestibule* can be seen between the *cella* and the stairway or in between the *cella* and the columns along the facade. On the base of this information it is plausible to suggest the idea that the *pronaos* would have been deeper in this particular plan layout. It would have been irregular for the *cella* to have been partitioned; which is clear, except for the examples of Pompeii and the Iseum in Cyrene. Meanwhile, in many plans we can clearly identify the foundation walls of the altar. For instance, in Ostia, Gighithis, Miletus and Ras el-Soda. The distinctive feature of this last example is the cult space's similarity to the layout of the Mythreum's *cella*. Otto Sosztarits' observation that the shrine had altars or statue niches all along the length of the temple, similarly to its counterpart in Pompeii, cannot be determined based on the examples we have at our disposal.

Exploring possible meanings of unexplained elements of the plan

The role of the cross walls in the cella

Hajnóczi's reconstruction does not answer the question of the

10 In HAJNÓCZY-SZENTLÉLEKY 1964, 134. Szentléleky compares the differrence of widths of the *pronaos* and the *cella* from the Szombathely structure to that of the Iseum and Serapeum structures in Pompeii and Rome. Available plans of these sites (including those of fig. 2) do not support this supposition. Neither does it a plan of the Iseum et Serapeum in the Campus Martius in Rome, whiche dates from the middle of the 19th century. The Iseum et Serapeum plan from Rome and other plans shown are taken from WILD 1984, 1739-1851.

architectural or structural role of the cross-walls in the *cella*. The remains of four cross-walls were found in the reconstruction under the *cella* in the *podium*. The second wall from the entrance is particularly strong and nearly three times the average wall thickness. It was clearly designed to have the same thickness as the main walls and it was altered later on. The two casings that flank the middle layer of the three-layer wall are not connected in the middle and were not built at the same time. Similarly, the outer walls and the walls surrounding the *cella* are independent from each other. Therefore, these walls must have been constructed afterwards to strengthen the centre wall. From the type technology and the kind of stone employed in the walls we can infer that these reinforcements were made approximately at the same time as the original walls. The depth of the walls is unknown but it could be determined using proper methods and then we could know if the middle wall and the other walls reach to the same depth. Cross-walls 3 and 4 are structurally identical with wall 1, but the northwestern part of wall 4 is missing, since it was damaged during World War II. According to the excavators' testimony and the excavation's filed notes the remains of another larger wall emerged in front of the main entrance wall or the *pronaos*. The top of this wall runs into the remains of the original ambulatory level. This data was used as the basis for determining the height of the podium. It is not possible to study the original condition of this arrangement, because part of the wall has been covered with the reinforced concrete structure.

The position of the stairs from the standpoint of function, analogy and archaeological data

The need to have enough room for the performance of liturgy is the main reason to question Hajnóczi's notion of the stairwell's location in the *pronaos*. The reference to the so called scribe's rooms cast even more doubt. Surveying the previous analogous examples, we have not found any examples of the stairs "biting" into the *pronaos*. However, it is not uncommon to find this arrangement in Roman temple architecture. For instance in those places where the stairwell is located in between two rows of columns, as in the temple in the harbour of Xanthen, the pronaos offers proportionally larger space than in Szombathely. Available data also casts doubts on the position and size of the stairs leading up to the *podium*. It seems to be a fact that, according to the current understanding, the *pronaos*' walls construction is of a lesser quality than the foundation walls excavated around the *cella*.

***The cornice fragments uncovered by the excavation and a new understanding of their position from analogue examples – Attic in the three-part cornice*.**

The difference in width between the *pronaos* and the *cella* came to light from the reconstruction of the frieze. I am unaware of other examples of this formation in Roman architecture. A new examination of the relief fragments calls into question the reconstruction of the frieze, and therefore it seems unavoidable to examine directly the marble fragments. Architectural discussions cannot neither confirm nor disprove the current reconstruction of the frieze. In form the reconstruction of the cornice is completely logical. Since a temple of Thebes was used as an analogy for the reconstruction of the cornice[11] (Pl. 25/1) there is no evidence that the cornice would have been closed with an attic space. It is also doubtful that such ill solution would have been employed on such significant building. The final analysis will be the task of archaeologists and art historians.

Hajnóczi determined the existence of an attic on the basis of a single small limestone fragment of a top section. The temple's counterpart in Thebes does not have such a feature. As we indicated regular format would rule out that a narrower *pronaos tympanic* roof would connect to the "bun-like" form of the *cella*. Therefore, it is definitely necessary to reconsider this aspect of the reconstruction. Given that this is a unique solution, its use could be verified through analogies found in Roman

11 HEGEDUS 1998.

architecture. Nevertheless, if we doubt the position of the relief's fragments on the frieze and we eliminate the difference of width between the *pronaos* and the *cella*, it is unnecessary to envision the building's format with the known roof-formation. Then it is plausible to imagine the frieze at the width of the *cella* and the building's reconstruction becomes a regular podium temple. The ominous top section fragment cannot be conclusively proven to be part of a horizontal or a slanted cornice, since it does not have a row of cranellations to attest its position. We can rightly suggest that this feature could also have been part of the slanted cornice. Then we can reconstruct a traditional Roman temple roof. The *pronaos* reconstruction that we have obtained reinforces the notion that this would have been the format used.

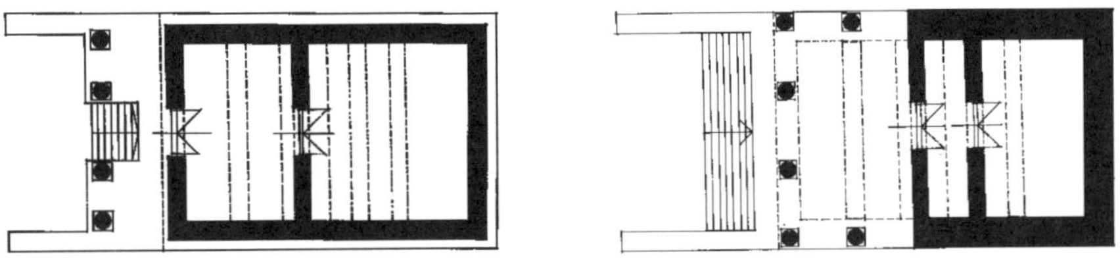

Fig. 7. The reconstructions of the plan by Hajnóczi and Mezős

A new concept for the Isis temple's reconstruction

Archaeological data, the space's functional analysis and an examination of the materials found in situ make the basis for formulating a new theory of the reconstruction of the site. As part of our proposal for a simplified version of the podium temple complex, we can assume it had walls that were constructed evenly. Instead of Hajnóczi's reconstruction of the *pronaos* we envision the entrance stairway constructed with a width that would essentially make it an extension of the building's side walls. The number of steps and the depth of the entrance agree with Hajnóczi's conclusions. In preparing our new proposal (fig. 7) we also raised the question of the cross walls under the *cella*. We thus propose that the columns of the *tetrastyle* facade (fig. 8) could have stood at the base of the main facade's wall. The second cross wall is three times thicker than the average walls and it could have been designed to bear a greater load, being the place of the facade wall. The pronaos then would have the depth of two spaces in

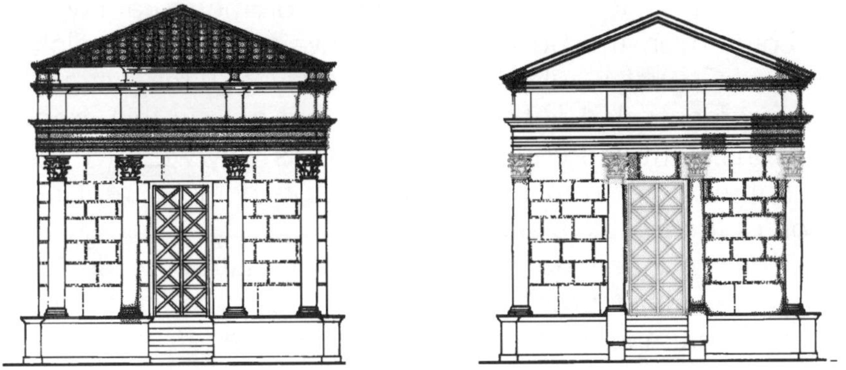

Fig. 8. The rekonsrtuctions of the main facade by Hajnóczi and Mezős

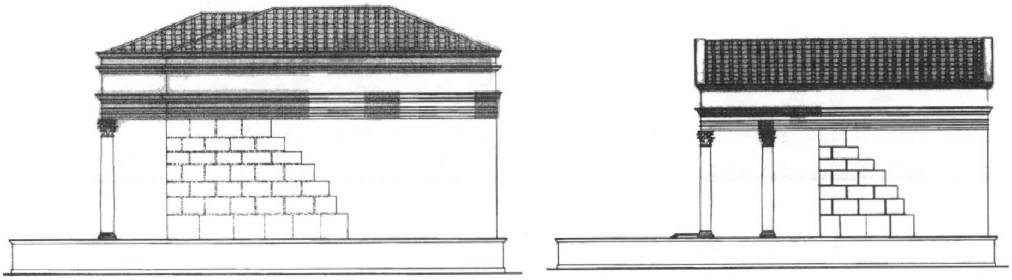

Fig. 9. Side-elevation by Hajnóczi and Mezős

between the rows of columns. Half of it would have been dissected by cross walls (fig. 9). In the position of this cross wall, there would have been columns 5 and 6, similar in form to that of the temple in Thebes that we used for an analogy. The first of the two cross walls behind the facade wall would be a partition wall, based on our analogous examples, the second could possibly have been the foundation wall of an altar for storing objects related to the cult's ritual.

In the proposed six-column version of the reconstruction, the axial distance of the outer columns is commensurate with that of the column on the building's lengthways side. The distance from column to wall would be approximately 2.80 ms. István Tóth in the aforementioned publication said that data in this area has been inconsistent and contradictory. This is why we have strictly used confirm data from the excavation in doing a real reconstruction of the temple's side facade.

The proposed new theoretical reconstruction is logical but requires more proof in order to prove conclusively the *pronaos*' 6 columns design. Naturally the pure *tetrastyle* facade order is also possible, but in this case the function of the unusually thick walls in the *cella* would not be possible to explain. It is also plausible that the foundation walls were, as in case of Egyptian temples of Isis, side walls of a sunken treasury under the ambulatory.

The Isis temple's current condition

The reconstruction unveiled in 1963 was done in the spirit of that decade modernism, with simplified formats, using reinforced concrete to display the various parts of the *tetrastyle* temple. The objective, consciously and publicly, was not to depict an original reconstruction. The reconstruction of the facade was based on the cornice fragments done in a sort of reinforced concrete installation. Almost as an *anastylosis*, it incorporated the original marble fragments. The year after the reconstruction was done, it was accepted as the first successful demonstration of the principles of the Venice Charter and it was displayed prominently on a poster. Since then the signs of physical deterioration can clearly be seen on the reinforced concrete structure. The construction carried out in a sub-standard manner, is damaged by corrosion of the reinforcements and by cracking of the concrete. Their is no insulation on the reinforced concrete covering above the cornice, meaning that any precipitation goes directly onto the marble reliefs. A scientific examination of the original stone material has not been done and only observations based on similar examples have been used. It is likely that direct contact between the limestone and the concrete is the cause for deterioration of the stone.

It is urgent to initiate a procedure to save these extremely valuable reliefs. There is no other solution but to take the Roman stones out of the concrete structure and substitute them with suitable copies. Simultaneously, or shortly afterwards, a complete excavation of the total site of the temenos is needed. We hope that finding new information or evidence will help us to answer conclusively the questions pertaining the uncertainties in the reconstruction.

Tamás Mezős
Technical University
Budapest

A Ring Decorated with the Bust of Sarapis in the Collection of the Hungarian National Museum

The Roman Collection of the Hungarian National Museum holds a gilded bronze ring which is decorated with the plastic bust of Sarapis. It is put on an unadorned ring that has a hemicircular cross-section (Pl. 27/1-2). The ring was taken into the inventory in the 1950s together with other objects from an unknown provenance. (Its inventory number is 54.33.111. Its size: the diameter of the ring: 20 mm, the height of the bust: 18 mm.) The circumstances under which it was found are not known, however it is probable that it comes from the Carpathian Basin (Pannonia or Dacia).

The incomplete hoop of the ring is widened out into an oval shape under the bust which is moulded together with the ring in order to provide a better support and bearing on a larger surface. The portrayal of the Sarapis bust, which is placed in a tangential position to the hoop of the ring and is closed in an arch at the bottom, is identical to the usual Hellenistic iconography of Sarapis busts.

The deity is wearing a *chiton* running down from his shoulders with vertical pleats and arranged in a wedge shape under his neck. His neck is covered with his fairly short beard parted in the middle, arranged in tiny curls. His slightly longish head, which can be put into an isosceles trapezium frame, is set off by his almost shoulder-long wavy hair falling into two parallel plastic curls at the front, while at the back it is flat, arranged in vertical curls following the shape of the head. There are five locks falling into his forehead. The main features of the exquisitely modelled face are the emphasized round eye-sockets and the strongly highlighted symmetrical round shaping of the area of the lips defined by the moustache and his beard. His remarkably small nose is incomplete.The *kalathos,* slanting backwards on the top of the head as a symbol of fertility is decorated with a plastic wreath pattern running around it horizontally. The Sarapis bust is level with the ring, being parallel with it so that the wearer can see it from a convenient view holding his hand in front of him.

The busts of Alexandrian gods were popular decorative motifs – we can see them on furniture or lamps. A plastic Sarapis bust is fastened to the strap handle of a terracotta lamp from Mainz which is similarly to the Sarapis ring at the National Museum (Pl. 27/1).[1] The bust was erroneously identified with Sabasios. Other examples of busts can be found on bracelets and on a special type of spiral rings on which the busts of Sarapis and Isis are fixed on either end. Such golden rings are in the British Museum (Fig. 1b)[2] and the Antiquarium in Berlin.[3]

There are more than 25 known examples of rings decorat-

[1] MENZEL 1969, 28 Nr. 90 Fig. 26, 7

[2] MARSHALL 1907, 41 Nr. 241 Fig. 9, Pl. 6.

[3] GREIFENHAGEN 1961, 87 Nr. 13, 97 Fig. 12.GRIMM 1969, 182; LECLANT 1971, 253 Note. 43; HORNBOSTEL 1973, 279 Note. 2. Pieces that have been discovered: MARSHALL 1907 41 Nr. 239 Pl. VI=Iside 286, IV. 331; MARSHALL 1907, 41 Nr. 240 Pl VI; MARSHALL 1907, 178 Nr. 1122 Pl. XXVII; MARSHALL 1907, 204 Nr. 1298 Pl. XXXI=Iside 285 IV 330; MARSHALL 1907, 204 Nr. 1301 Pl. XX; MARSHALL 1907, 204 Nr. 1302 Pl. XXXI; MARSHALL 1907, 204 Nr. 1298 Pl. XXXI= Iside 285 IV 330; MENZEL 1964 Nr. 142=HORNBOSTEL 1973, Pl. CLXXV 282. Egypt: KARO 1901, 212 Fig. 11=EDGAR 1904, 1 Nr. 27.631; EDGAR 1904, 65 Nr. 27.891 Pl. XVII; Alexandria: SCHREIBER 1894, 308 Nr. 40 Fig. 40; Dalmatia (Turbe): KATER-SIBBES 1973, Nr. 907=CLERC - LECLANT 1994, 678 Nr. 88 c; Den Haag: HORNBOSTEL 1973, 279 Note. 2, Taf. CLXXV Fig. 283; Dieburg: HENKEL 1913, 97 Nr. 1038 Pl. 41=GRIMM 1969, 203 Nr. 113 Taf. 35, 2-4; Monastir (Sardinia): GAUCKLER - MERLIN 1910, 132 Nr. 125 Pl. LVIII, 8; Paphos: KARAGEORGHIS 1959, 346 Fig. 13=JHS 1958, 29 Fig. 9; Royan (Charente-Maritime): LABROUSSE 1952, 93-95 Fig 9=LECLANT 1961, 405 Nr. 18a=KATER-SIBBES 1973, 153 Nr. 812; Zugmantel: GRIMM 1969, 181-182 Note 23-24=CLERC-LECLANT 1994, 678 Nr. 88a.

ed with the plastic busts of Alexandrian gods, to which the Sarapis ring of the Hungarian National Museum belongs, but the majority of them are unfortunately unpublished.[4] All of the rings found and published depict Sarapis, in three cases (a ring from Egypt (Fig. 2/4)[5], one from Royan (Fig. 2/3)[6],and one held in the British Museum (Fig. 3/6)[7]) in the form of Sarapis-Ammon respectively. On the top of the head of Sarapis, with the exception of three rings decorated with Sarapis busts wearing the crown *atef,*[8] we can find the *kalathos*. Together with Sarapis, his *paredros*, Isis is also placed on another ring held in the British Museum (Pl. 28)[9], and there is only one ring known with a single Isis figure. This ring in the British Museum from the Hamilton Collection is a variant of the type since Isis, who is holding a *sistrum* and a jug in her hands, is in full figure.[10]

The general characteristics of the Sarapis rings (relevant to the majority of them) are the following:

1. The narrow hoops have a hemicircular cross-section. The hoop widens out into an oval shape and flattens out under the plastic busts. The ring from Monastir held in the Musée Alaoui is, however, rather thick. It has an oval cross-section and does not widen out under the tiny Sarapis bust (Fig. 4/2)[11]

2. In most cases the busts are not soldered to the hoop, but

[4] For more Sarapis rings see: Further pieces without published pictures: Victoria and Albert Museum, London Waterton Collection Inv. Nr.: 534-1871; Brooklyn Museum Inv. Nr.: 67, 2; Collezione L. Sanlini, Vendita Roma 24. Inv. Nr.: 26.4.1899 Nr. 191; University College, London, Petrie Collection: CASTIGLIONE 1978, 230 Nr. 53

[5] KATER-SIBBES 1973, 47 Nr. 290 Pl. V

[6] LABROUSSE 1952, 93-95 Fig 9=LECLANT 1961, 405 Nr. 18a

[7] MARSHALL 1907, 204 Nr. 1301 Pl. XX.

[8] HORNBOSTEL 1978, 510 Note 30= CASTIGLIONE 1978, 230 Nr. 53; MARSHALL 1907, 1299=CASTIGLIONE 1978, 230 Nr. 54 (Pl. IV. 5); MARSHALL 1301=CASTIGLIONE 1978, 230 Nr. 55 (Pl. IV. 6).

[9] MARSHALL 1907, 204 Nr. 1298 Pl. XXXI=Iside 285 IV 330.

[10] MARSHALL 1907, 41 Nr. 240 Pl VI.

[11] GAUCKLER - MERLIN 1910, 132 Nr. 125 Pl. LVIII, 8.

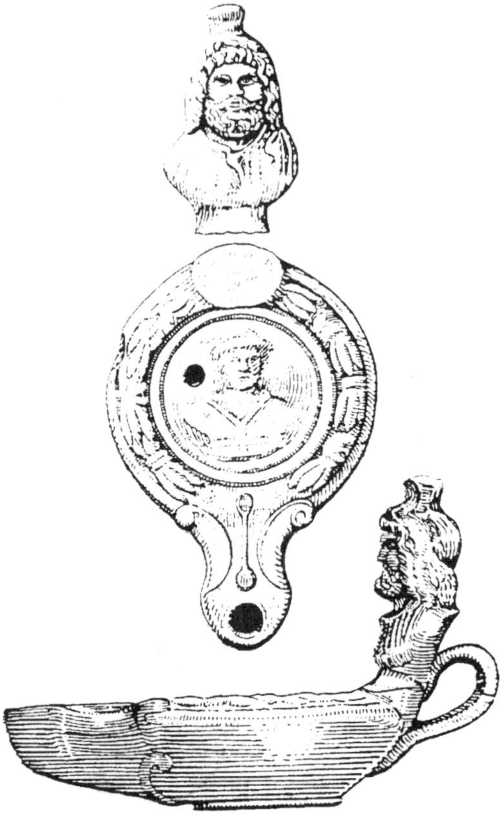

1

2

Fig. 1a. Lamp decorated with Sarapis bust from Mainz 1b. Spiral ring with the busts of Isis and Sarapis in the British Museum

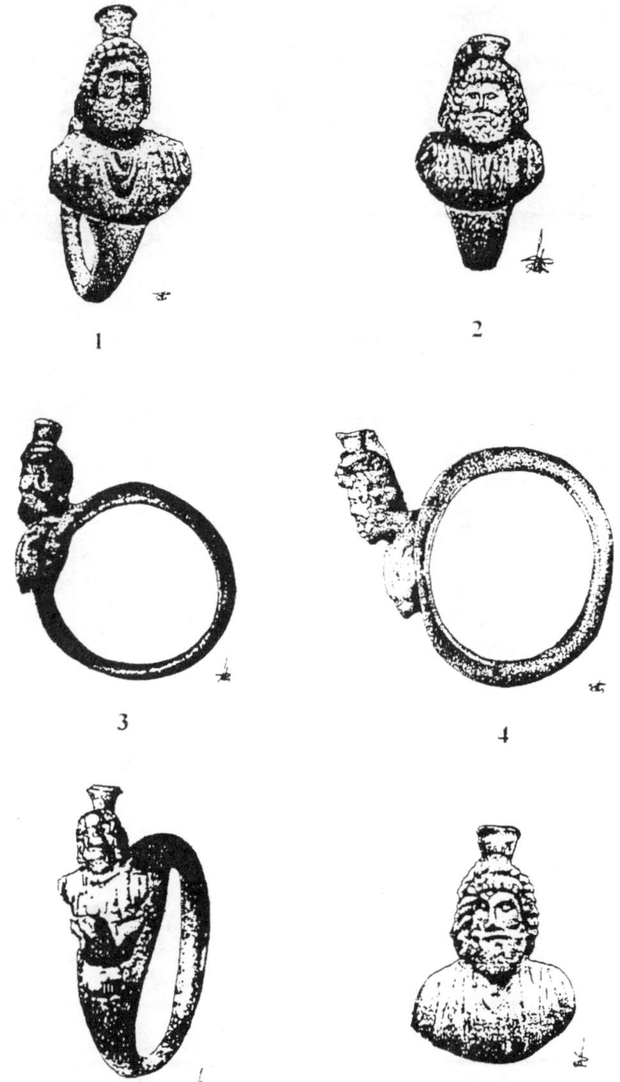

Fig. 2. Rings decorated with plastic Sarapis busts. 1. in Hannover, Kestner Museum: MENZEL 1964 Nr. 142; 2. in the British Museum: MARSHALL 1907, 178 Nr. 1122; 3. from Royan: LABROUSSE 1952, 93-95; 4. from Egypt: KATER-SIBBES 1973, 47 Nr. 290; 5. in the British Museum: MARSHALL 1907, 204 Nr. 1302; 6. in den Haag, Koninklijk Kabinet van Munten: HORNBOSTEL 1973, 279, Anm. 2.

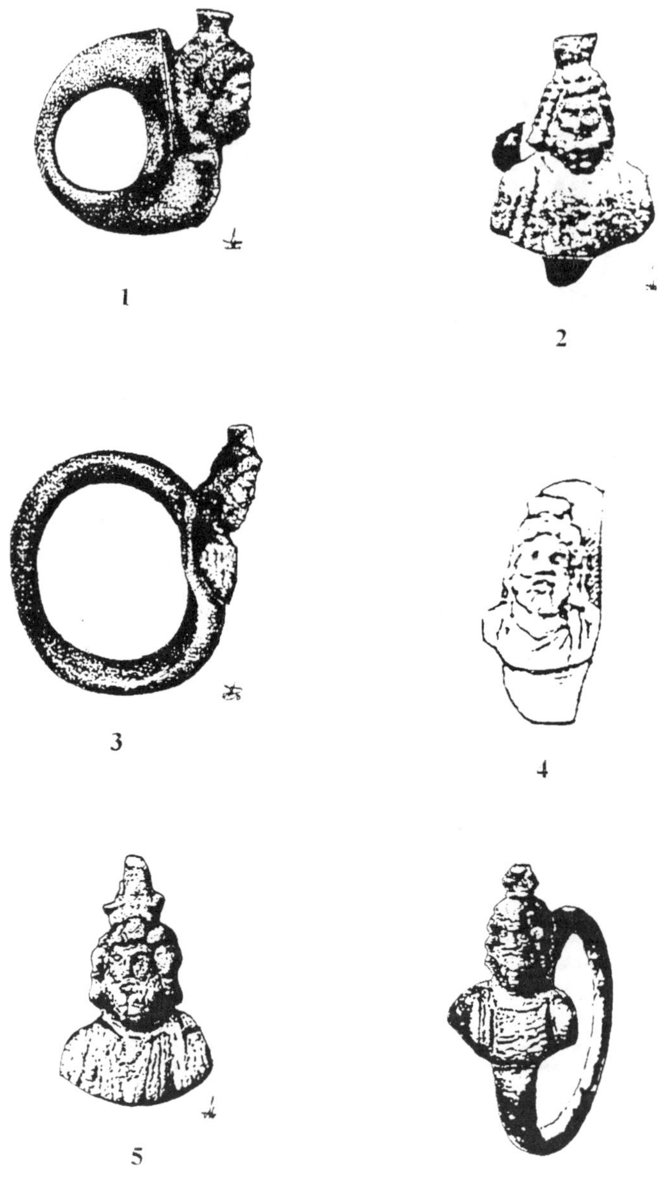

1 2 3 4 5 6

Fig. 3. Rings decorated with plastic Sarapis busts. 1. in Walters Art Gallery: HILL 1946, 67; 2. from Egypt: EDGAR 1904, 65; 3. Karo 1901, 212; 4. from Alexandria: SCHREIBER 1894, 308 Nr. 40; 5. in the British Museum: MARSHALL 1907, 204 Nr. 1299; 6. MARSHALL 1907, 204 Nr. 1301.

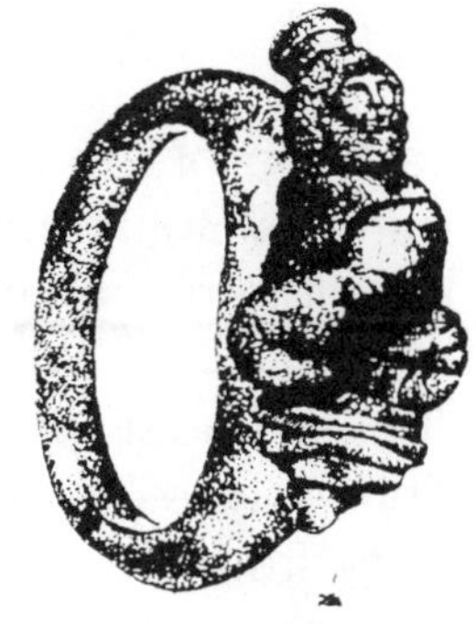

1

2

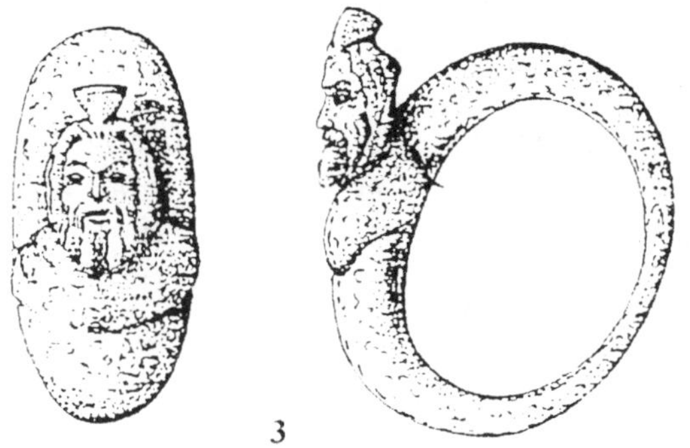

3

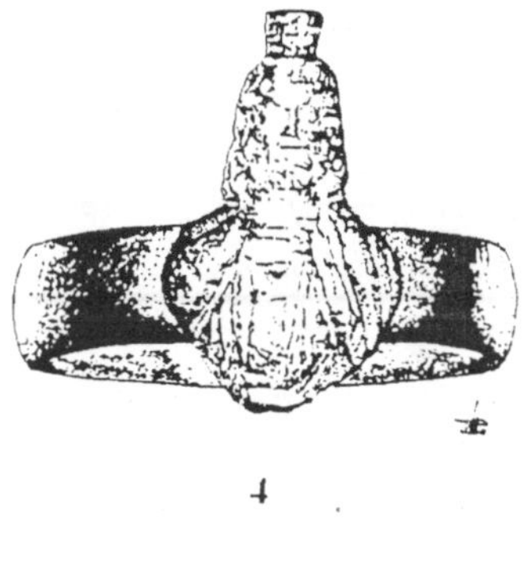

4

Fig. 4. Rings decorated with plastic Sarapis busts: 1. from Paphos: KARGEORGHIS 1959, 346; 2. from Monastir: GAUCKLER - MERLIN 1910, 132 Nr. 125; 3. From Zugmantel: GRIMM 1969, 181-182 Nr. 82; 4. in the British Museum: MARSHALL 1907, 41 Nr. 239.

are moulded together with the ring. We can find separately fastened busts on rings from Zugmantel (Fig. 4/3)[12], den Haag (Fig. 2/6)[13] and Egypt (Fig. 2/4)[14].

3. The Sarapis busts are level with the hoop of the ring. The only exception is the golden Sarapis ring in the British Museum, on which the bust is perpendicular to the ribbon-like ring (Fig. 4/4)[15]

4. Unlike a number of bronze or marble busts, under the Sarapis busts there are no globes or eagles. The only exception is a bronze ring which is in the British Museum from the Castellani Collection, where the bust is cast upon the back of an eagle with its wings spread out, standing on a small base (Zeus-Sarapis) (Fig. 2/5)[16]. On the ring from Paphos, the round-headed Sarapis bust is supported by a snake coiled up in the shape of a horizontal figure eight (Fig. 4/1)[17].

5. The rings are usually 19-24 mm in diameter, the height of the busts is between 15-20 mm. A silver Sarapis ring from the British Museum has the largest diameter of 29 mm (Fig. 2/2), while the largest Sarapis bust is 27mm high and it came to light in Egypt (Fig. 3/2).

Most of the rings are made of bronze[18] or gilded bronze and a few of them are made of silver[19] or gold[20].The ring from den Haag is made of gold, while the bust on it is made of silver

[12] GRIMM 1969, 181-182 Fig. 23-24.

[13] HORNBOSTEL 1973, 279 Note 2, Taf. CLXXV Abb 283.

[14] KATER-SIBBES 1973, 47 Nr. 290 Pl. V.

[15] MARSHALL 1907 41 Nr. 239 Pl. VI=Iside 286, IV. 331.

[16] MARSHALL 1907, 204 Nr. 1302 Pl. XXXI.

[17] KARAGEORGHIS 1959, 346 Fig. 13=JHS 1958, 29 Fig. 9.

[18] EDGAR 1904, 65 Nr. 27.891 Pl. XVII; HENKEL 1913, 97 Nr. 1038 Taf. 41=GRIMM 1969, 203 Nr. 113 Pl. 35, 2-4.

[19] MARSHALL 1907, 178 Nr. 1122 Pl. XXVII; KATER-SIBBES 1973, Nr. 907=CLERC - LECLANT 1994, 678 Nr. 88 c.

[20] From Egypt: KATER-SIBBES 1973, 47 Nr. 290 Pl. V; from Paphos: KARAGEORGHIS 1959, 346 Fig. 13=JHS 1958, 29 Fig. 9; unknown provenance: MARSHALL 1907 41 Nr. 239 Pl. VI=Iside 286, IV. 331.

(Fig. 2/6)[21]. L. Castiglione mentions an iron ring decorated with the plastic busts of Isis and Sarapis from the Petrie Collection (University College, London)[22].

The place where the rings were discovered is in most cases unknown, but in a few cases the place of origin can be identified exactly or approximately. We know of five Sarapis rings from Egypt (one of them is undoubtedly from Alexandria)[23], one from Africa[24], one from Greece[25], one from Cyprus (Paphos)[26], one from Dalmatia[27], possibly two from the Danube Region[28], three from Germania[29] and one from Gaul[30]. The habit of wearing a ring decorated with a Sarapis bust – just like wearing rings with gems and cameos with representations of the Alexandrian gods – was quite popular on the whole territory of the Roman Empire including the northern provinces (Gaul, Germania, the Danube Region) as well.[31]

The type of rings decorated with the plastic busts of gods is

21 HORNBOSTEL 1973, 279 Note 2, Pl. CLXXV Abb 283.

22 CASTIGLIONE 1978, 230 Nr. 53.

23 KARO 1901, 212 Fig. 11=EDGAR 1904, 1 Nr. 27.631; EDGAR 1904, 65 Nr. 27.891 Pl. XVII; SCHREIBER 1894, 308 Nr. 40 Fig. 40; KATER-SIBBES 1973, 47 Nr. 290 Pl. V; KATER-SIBBES 1973, 25 Nr. 135.

24 Monastir: GAUCKLER - MERLIN 1910, 132 Nr. 125 Pl. LVIII, 8.

25 SEGALL 1938, Nr. 146/7 Pl. 39.

26 Paphos: KARAGEORGHIS 1959, 346 Fig. 13=JHS 1958, 29 Fig. 9.

27 From Turbe in the Sarajevo Museum: KATER-SIBBES 1973, Nr. 907=CLERC - LECLANT 1994, 678 Nr. 88 c.

28 DREXLER 1890, 96; MNM Inv. Nr.: 54.33.111.

29 Den Haag: HORNBOSTEL 1973, 279 Anm. 2, Taf. CLXXV Abb 283; Dieburg: HENKEL 1913, 97 Nr. 1038 Taf. 41=GRIMM 1969, 203 Nr. 113 Taf. 35, 2-4; Zugmantel: GRIMM 1969, 181-182 Fig. 23-24=CLERC-LECLANT 1994, 678 Nr. 88a.

30 Royan (Charente-Maritime): LABROUSSE 1952, 93-95 Fig 9=LECLANT 1961, 405 Nr. 18a.

31 The rings decorated with the Alexandrian gods from the Northern provinces of Rome (a selection): From: Germania: GRIMM 1969, 119 Nr. 2

not of Roman origin. The earliest Sarapis ring, held in the Walters Art Gallery in Baltimore belongs to the Hellenistic type due to its shape, which is narrow at the bottom, gradually widening out towards the bezel of the ring. Its peculiarity is that the relief-like Sarapis bust of the large and heavy ring is put on the large, oval bezel (Fig. 3/1)[32]. Thus, the origins of the rings decorated with plastic busts are probably Hellenistic on the evidence of the areas where the rings were used and the represented deity (Sarapis). The origins are found in Egypt during the Ptolemaic era, and more closely in Alexandria. Most Sarapis rings are known from Egypt, from the Imperial period.[33] Probably some rings from unknown provenance are from Egypt, but as far as the place of discovery is concerned, only one of them can be related to a single settlement, namely Alexandria.[34] Probably here, in the centre of the Sarapis faith the production of this particular type of rings started. Since they were popular, the form must have been prevailing from here into the European provinces of the Roman Empire.

The rings must have been produced in different workshops as they were made of different materials, using different procedures and the busts were modelled in a distinctive way. There is only one of these workshops – probably the most significant – from which we know that more than one, that is, five pieces belong to a single series. Four of these similarly shaped and modelled rings have the same distinctive features as the Sarapis ring owned by

(Harpokrates); GRIMM 1969, 148 Nr. 31 (Isis-Fortuna, Sarapis, Hygieia); GRIMM 1969 120-121 Nr. 3 Pl. 35, 7 (Sarapis); from Pannonia: GESZTELYI 2000, 52 Nr. 87 (Sarapis); an unpublished ring with Isis and Serapis: MNM 81.1882.1; the portrait of Sarapis on the *cameo* of a golden ring from the graveyard in Intercisa: GESZTELYI 1998,141 Nr. 38.

32 HILL 1946, 67 Nr. 5 Fig. 6.

33 KARO 1901, 212 Fig. 11=EDGAR 1904, 1 Nr. 27.631; EDGAR 1904, 65 Nr. 27.891 Pl. XVII; SCHREIBER 1894, 308 Nr. 40 Fig. 40; KATER-SIBBES 1973, 47 Nr. 290 Pl. V; KATER-SIBBES 1973, 25 Nr. 135.34 SCHREIBER 1894, 308 Nr. 40 Fig. 40.

34 SCHREIBER 1894, 308 Nr. 40, Fig. 40.

the Hungarian National Museum (Pl. 27/1-2). The bronze ring, which was found in Dieburg and was lost during the Second World War, was probably made in the same workshop. The regular round shape of the eye-sockets and the lips were clearly noticeable on the worn surface of this ring (Pl. 27/3-4)[35]. Another ring, held by the Cairo Museum (Fig. 3/3)[36] and the Sarapis-Ammon ring from Royan (Fig. 2/3)[37], probably belong to the same series due to their similar characteristics. This workshop probably moulded rings not only with Sarapis or Sarapis-Ammon busts, but also with that of Isis. Such is the gilded bronze ring of unknown origin now in the British Museum, decorated with the busts of Sarapis and Isis (Pl. 28)[38] on which the modelling of Sarapis is identical to that of the plastic bust on the ring from the Hungarian National Museum.

Based on the evidence of the ring found in Egypt, we can assume that the most productive and most important workshop was in Egypt, more precisely in Alexandria, a centre of bronze and silver-smiths[39] and also the regional centre of the Sarapis faith. The products of this workshop turned up in the northern Roman provinces as well. Consequently it must have continued production during the Imperial period. Since all the plastic busts depict Sarapis (in two cases together with Isis), the rings in question can be dated to the period when Isis was losing her importance and Sarapis, who enjoyed the support of the imperial house, became the central figure of Egyptian cults. This shift of emphasis happened all over the empire in the Severian-period. Also, Günter Grimm dated the Dieburg ring of the Alexandrian workshop to this very period.[40]

[35] HENKEL 1913, 97 Nr. 1038 Pl. 41=GRIMM 1969, 203 Nr. 113 Pl. 35, 2-4.

[36] KARO 1901, 212 Fig. 11=EDGAR 1904, 1 Nr. 27.631; EDGAR 1904, 65 Nr. 27.891 Pl. XVII.

[37] LABROUSSE 1952, 93-95 Fig 9=LECLANT 1961, 405 Nr. 18a=KATER-SIBBES 1973, 153 Nr. 812.

[38] MARSHALL 1907, 204 Nr. 1298 Pl. XXXI=Iside 285 IV 330.

[39] SCHREIBER 1884, 271-480; DREXEL 1909,176-235.

[40] GRIMM 1969, 203. His dating was accepted by CLERC-LECLANT 1994, 678 Nr. 88 b.

The ring stands for the symbol of relationships, the feeling of togetherness, belonging to someone either in a social class, for example, engagement or wedding rings[41] or rings which demonstrate its bearer's close, friendly relationship to the imperial family[42] or his social standing[43], or rings as insignia of military ranks[44], or in religious, magical relations[45].

The establishment and maintenance of a close and permanent relationship between man and the deities is most effectively expressed through a ring as the closest and most personal object that we wear. The existence of these close relationships were justified with the inscriptions dedicated to gods[46] and also the portraits and attributes of gods as decorations found on several rings[47].Thus the rings decorated with the Sarapis bust can be strongly connected to the cult of Sarapis, and the owners might have been among the believers of Sarapis. From this point of view it is not immaterial that the text in which Aulus Gellius refers to the *Liber Aegyptiacus* by Apion gives us an explanation of why the Greeks and the Romans wore their rings on the fourth finger of the left hand. That is, in Egypt while dissecting human corpses a tiny nerve had been found which connected this finger with the heart, as the seat of the soul (Gellius, Noctes Atticae, X, 10.)[48].

The frequent occurrence of the Sarapis images on rings is due to the humane, helpful, supportive attitude of the deity - just like Isis – who listened to people's everyday problems, gave answers to their requests and brought luck and happiness as

41 DÖLGER 1936, 188-200; BORHY 1997, 99-100.

42 TÓTH 1983-1984, 315-336.

43 ALFÖLDI 1952, 27; FOURLAS 1971, 76-82; KOLB 1977, 247, 249.

44 GESZTELYI 1998, 141 Nr. 38.

45 TÓTH 1980, 94; TÓTH 1985, 21-25.

46 Summarised in: TÓTH 1980, 94-95. Another votive ring: PLANCK 1991, 57-58, 60 Fig. 83.

47 GANSCHINIETZ 1914, 835.

48 GANSCHINIETZ 1914, 837.

well. Sarapis, together with Isis, guaranteed fertility and affluence and controlled the unforeseeable fate. Therefore their believers, according to their faith, were able to overcome their destinies with the help of their gods[49]. These philantropic attributes transformed Sarapis and Isis into healing figures who kept illness away, insomuch as Sarapis became one of the three most important healing gods of the Roman Empire by the 3rd century AD.[50] Wearing the images of Isis and Sarapis, primarily on rings[51] and fibulas[52], guaranteed the intimate connection and the constant support of gods for their believers in problems of everyday life and health, and rendered help in achieving their further aims. The following formula can be read on a gem: πάντα νικά ὁ Σάραπις.[53] The φιλοσαρὰπις Caracalla was wearing the image of the invincible Sarapis who guaranteed a certain victory, on his breastplate and he went to the battlefield wearing the image of his god.[54]

In connection with rings decorated with plastic busts, Günter Grimm raised the question of the apotropaic qualities of the rings bearing the images of Alexandrian gods.[55] The general meaning of rings as closed circles is that they are like magic circles protecting their owners from all evil spirits and powers.[56] It is quite possible, based on the Sarapis rings, that this feature of the ring, together with the protective qualities of Alexandrian gods, kept evil spirits away from their believers. It is in this way they truly possessed the

49 HÖLBL 1983, 98-99.

50 ROEDER 1920, 2422; BECHER 1970, 228-233; HORNBOSTEL 1973, 22; HÖLBL 1983, 99; HÖLBL 1984, 871; MRÁV 2000, 85-86; Cass. Dio 78, 15, 2. The wearing of the ring and the recovery: GANSCHINIETZ 1914, 837-838.

51 MARSHALL 1914, 823-824; THOMAS 1968, 119-127.

52 KUZMOVÁ 1997, 81-82.

53 CIG IV 7044 vö.: ROEDER 1920, 2428.

54 HORNBOSTEL 1973, 241 Note 2, 380 Note 4, Nr. 87 Taf. 3; HORNBOSTEL 1978, 515-516; MRÁV 2000, 87-88.

55 GRIMM 1969, 61; HÖFLER 1935, 62.

56 GANSCHINIETZ 1914, 836.

protective power in the mind of their owner.

Some rings decorated with the image of Sarapis come from a sepulchral background. A golden ring with a Sarapis cameo was unearthed from a grave in Intercisa (Dunaújváros),[57] just like the silver ring with Isis and Sarapis which was found in a graveyard in Sedeinga, Sudan in 1969.[58] It is not certain whether the rings were put into the grave because of the beliefs connected to Sarapis as the ruler of the other world[59] or because they were simply the personal jewels of the dead put into the grave.[60] It is also possible that both aspects mattered when burying the rings with their owners.

The rings of the Alexandrian workshop found in the northern provinces of the Roman Empire, like the one in the Hungarian National Museum, were probably not the result of commercial activities. Instead they were carried around in tourists' and pilgrims' travels to the main cultic places of the Alexandrian gods, which were quite usual in the Imperial period.

Zsolt Mráv
Hungarian National Museum
Budapest

Addition

Two other Sarapis-rings: F. Paolucci, Museo nazionale del Bargello. Reperto archeologici, Firenze 1994, 92-93.
Two falsified Sarapis-ring from the 19th century: G. Zahlmaas, Fingerringe und Gemmen. Ausstellungskataloge der prähistorischen Staatsammlung, /Bd. 11, München 1985, 55, nr. 80-81.

[57] GESZTELYI 1998, 141 Nr. 37.

[58] KATER-SIBBES 1973, 25 Nr. 135.

[59] LEHMAN-HAUPT 1909-1915, 362; ROEDER 1920, 2419-2421; BONNET 1952, 651; CUMONT 1959, 36-37; HÖLBL 1983, 98.

[60] Rings come to light from grave: MARQUART 1879, 337, 3. By tradition in the Roman afterlife the deceased must not wear a ring: Plin. Nat. Hist. XXXIII 27; GANSCHINIETZ 1914, 837.

Annex

	Prove-nance	Material of the rings	Diameter of the hoop	The high of the Sarapis busts	Weight
1. Pl. I. 1-2; The Sarapis ring in the collection of Hung. Nat. Mus. (MNM Inv. Nr.: 54.33.111)	Unknown	Gilded bronze	20 mm	18 mm	
2. Pl. III. 1; GRIMM 1969, 203 Nr. 113 Taf. 35, 2-4.	Dieburg	Bronze	20 mm	18 mm	
3. Pl. III. 2; MARSHALL 1907, 204 Nr. 1298 PL. XXXI=Iside 285 IV 330.	Unknown	Gilded bronze	19 mm	13 mm	
4. Pl. IV. 1; HILL 1946, 67 Nr. 5 Fig. 6.	Unknown	Gilded bronze		The high of the bezel: 40 mm	
5. Pl. IV. 2; EDGAR 1904, 65 Nr. 27891	Egypt	Bronze	24 mm	27 mm	
6. Pl. IV. 3; KARO 1901, 212 Fig. 11	Egypt	Gilded bronze			
7. Pl. IV. 4; SCHREIBER 1894, 308 Nr. 40 Fig. 40	Alexandria	Gold	12,5 mm		
8. Pl. IV. 5; MARSHALL 1907, 204 Nr. 1299	Unknown	Gilded bronze	22 mm		
9. Pl. IV. 6; MARSHALL 1907, 204 Nr. 1301.	Unknown	Gilded bronze	20 mm		
10. Pl. V. 1; MENZEL 1964, Nr. 142; HORNBOSTEL 1973, 279, Taf. CLXXV Nr. 282.	Unknown	Bronze			
11. Pl. V. 2; MARSHALL 1907, 178 Nr. 1122 PL. XXVIII.	Unknown	Silver	29 mm	15 mm	3,36 gr

12. Pl. V. 3; LABROUSSE 1952, 93-95	Royan (Charente -Maritime)	Bronze	20 mm	20 mm	6 gr
13. Pl. V. 4; KATER-SIBBES 1973, 47 Nr. 290.	Egypt	Gold	23 mm	20 mm	
14. Pl. V. 5; MARSHALL 1907, 204 Nr. 1302.	Unknown	Bronze	21 mm		
15. Pl. V. 6; HORNBOSTEL 1973, 279 Anm 2. (Den Haag)	Unknown	gold hoop with silver Sarapis bust			
16. Pl. VI. 1; KARAGEORGHIS 1959, 346 Fig. 13.	Paphos	Gold	20 mm	23 mm	
17. Pl. VI. 2; GAUCKLER - MERLIN 1910, 132 Nr. 125.	Monastir	Gilded bronze			
18. Pl. VI. 3; GRIMM 1969, 181-182 Nr. 82 Abb. 23-24	Zugmantel	Gilded bronze	23 mm	15 mm	
19. Pl. VI. 4; MARSHALL 1907, 41 Nr. 239 PL. VI.	Unknown	Gold	18 mm	15 mm	3,3 gr.
20. KATER-SIBBES 1973, 179 Nr. 907	Turbe	Silver	24 mm		
21. KATER-SIBBES 1973, 25 Nr. 135	Sedeinga	Silver	15 mm	25 mm	
22. SEGALL 1938, Nr. 146/7 Taf. 39.					
23. Victoria and Albert Museum, London Waterton coll. Inv. Nr. 534-1871.					
24. Brooklyn Museum Inv. Nr. 67, 2.					
25. Coll. L. Sanlini, Vendita Roma Inv. Nr. 26.4.1899.					

Veränderungen des Hintergrundes der Ägyptischen Kulte in Pannonien im 1.-4. Jahrhundert

Der Isis-Kult wurde in der Kaiserzeit im Römischen Reich verbreitet. Ihr erstes gemeinsames Heiligtum mit Serapis ließ Kaiser Caligula im Jahre 38 in Rom errichten.[1] Die Göttin der tausend Gesichter und tausend Namen verlor ihren alexandrinischen Grundcharakter auch während ihrer hellenistisch-römischen Angleichung nicht: mit dem Steuerruder in der Hand gewährte sie vor allem Reisenden oder Handelsleuten auf dem Meer Schutz vor Stürmen und Seeräubern. In dieser Gestalt verehrte sie nämlich die Hafenstadt Alexandria (Hafengöttin).[2]

"Bald wurde sie aber zur Schicksalsgöttin rein römischen Typs....als Lenkerin menschlicher Geschicke hält sie das Steuerruder in der Rechten und als Spenderin von Glück und Reichtum das Füllhorn in der Linken."[3] So wurde sie später in den nördlihen Provinzen und im ganzen Reich als Isis-Fortuna verehrt.

Über Italien fand Isis sehr bald, schon in der zweiten Hälfte des 1. Jahrhunderts, auch den Weg nach Pannonien; damals noch ausschließlich als Schirmherrin des Handels. Die von den Schiffen auf das Festland verladenen Lieferungen genossen entlang der Bernsteinstraße weiterhin den Schutz der Göttin.

Eindeutig weist darauf die Inschrift eines vielzitierten Altarsteins von Savaria hin, den Tiberius Barbius Valens,

[1] VIDMAN 1981, 121.
[2] JOBST 1992, 77.
[3] JOBST 1992, 77; SZENTLÉLEKY 1960, 12.

Dekurion der Colonia Claudia Savaria, anfertigen ließ.[4] Der Familie Barbius gehörte während dieser Zeit, also in der zweiten Hälfte des 1. Jahrhunderts, das einflussreichste Geschäftshaus in den Städten an der Bernsteinstraße, dessen Filialen an allen Stationen der das Adriatische Meer mit der Donau verbindenden Nordsüdroute vertreten waren.[5] Nicht zufällig bekleidete ein Mitglied der Familie in Savaria, das unter den Flaviern im ungeteilten Pannonien zum Provinzsitz avancierte, den Dekurionenrang. Nach den Ergebnissen der jüngsten Nachgrabungen vermutet man, dass unter den Flaviern in Savaria das erste Isis-Heiligtum (Tf. 29/1) gebaut wurde. In der Umgebung sind auch aus dem 2. Jahrhundert mehrere vom *sacerdos provinciae* beziehungsweise den Priestern des Isis-Kultes gestiftete Altäre bekannt. Die vielköpfige Gemeinde pflegte den Kult der Göttin in Vertretung der italienischen Geschäftshäuser.[6]

An der nächsten Station der Bernsteinstraße, in **Scarbantia**, ist ein Altarstein in der Nachbarschaft des Capitolium-Tempels, an der Westseite des Forums, mit dem Gaius Pompinius(?) Publius Philinus, Libertus des Pomponius Severus, bekannt, wo dieser sein Gelübde gegenüber Isis und Bubastis erfüllte. Der Altar wurde etwa hundert Jahre später gestiftet als der oben erwähnte von Savaria.[7] Für den genauen Zeitpunkt bieten sich zwei Varianten an, (hierin ist die Forschung bislang geteilter Meinung). Die in der Inschrift vorkommenden Personen könnten zur Familie des G. Pomponius Bassus Terentianus gehört haben. Dieser hatte von 186 bis 187 das Amt des Statthalters der Provinz Pannonia Inferior inne. Der Patron des Libertus war also in Aquincum tätig. Wahrscheinlicher ist, dass der Stifter des Altares von Scarbantia, wegen der räumlichen Nähe, ein Familienangehöriger des als Statthalter in Carnuntum sitzenden T. Pomponius Protomachus ist. In diesem Fall übte der

4 CIL III, 4156, RIU I, 14. = MÓCSY 1959, 24.

5 MÓCSY 1959, 219.

6 BUÓCZ 1978, 83, 84, 86; TÓTH 1998, 130.

7 RIU I 164.

Patron sein Amt als Statthalter der Provinz Pannonia Superior in Carnuntum aus, und zwar während der Regierungszeit Caracallas. Nach 214 legt auch seine griechische Abstammung eher die letztere Vermutung nahe.[8]

Als dritte an der Bernsteinstraße wäre eine unter bedeutungsloseren Funden und Relikten herausragende Inschrift aus **Carnuntum** zu erwähnen. Die Aufzeichnungen bezüglich ihrer näheren Fundstelle sind widersprüchlich. (Einzelne Quellen zählen sie bei den Inschriften der Zivilstadt auf, andere geben das Militärbad des Legionslagers als Fundort an. Die "Serapi Conservatori et Isidi ceterisque Diis Deabusque immortalis ..." lautende Widmung des Altarsteins wurde für "P..A(elius)? S(...)cum Fabia Titiana uxore et Titiano filio ..." im ersten Drittel des 3. Jahrhunderts gestiftet, und zwar von seiner eigenen Familie.[9]

"Serapis war ein nach den Vorstellungen der neuen griechischen Landesherren hellenisierter ägyptischer Gott, als Gemahl der Isis trat er an die Stelle des altägyptischen Osiris. Wie Osiris war Serapis Herr der Toten, die er an seiner Unsterblichkeit teilhaben ließ. Für die Lebenden war er ein Heilgott im universalen Sinn. In Ägypten war sein Beiname soter, Heiland. Die Inschrift nennt ihn Conservator = Retter."[10]

In den auf die flavische Barbius-Inschrift von Savaria folgenden anderthalb Jahrhunderten wandelte sich der geistige Gehalt des Kultes; zur selben Zeit veränderten sich die "Kräfteverhältnisse" innerhalb Pannoniens, und auch die Hauptroute des Handels verlagerte sich.

Was den "inneren Gehalt" anbelangt, so ist die Veränderung im Großen und Ganzen in die Zeit der quadisch-markomannischen Kriege anzusetzen. In den Ungewissheit und Kriege bringenden Jahrzehnten löst die beschützende, geistige Erlösung versprechende, vertrautere Religiosität des Serapis-

8 FITZ 1994, 538 und 1024.

9 CIL III, 11157, CSIR I, S. 49, Nr. 442; VORBECK 1980, 19, Nr. 69.

10 JOBST 1992, 75-76, 79.

Kultes, die den Menschen näher kommt, die praktische Seite der Isis ab. Das deutet auch eine Bauinschrift aus Carnuntum an, die den Bau oder die Wiederherstellung eines Serapis-Isis-Heiligtums verewigte – nach den Namen der Konsuln zu urteilen im Jahre 213, zur Zeit Caracallas.[11] Daraus schloß man, dass, ähnlich dem Isis-Heiligtum von Savaria, auch in Carnuntum ein Tempel des Götterpaares gestanden haben muß, den eine grosse Gemeinde unterhielt.

Ausserdem hat die im Kult eingetretene Veränderung auch ökonomische Ursachen. Wie gesehen, war die Bernsteinstraße im 1. bis 2. Jahrhundert die Hauptschlagader der Warenversorgung in der Provinz. Von hier zweigten die Verkehrsverbindungen zu den übrigen Teilen der Provinz ab. Ende des 2. Jahrhunderts änderte sich diese Situation bekanntlich. Das an der Rhein-Donau-Grenze stationierte Heer wird nicht mehr aus Italien, sondern aus den westlichen Provinzen beliefert, so dass die Hauptverkehrsroute dem Limes folgt. Die meisten Investitionsgelder fließen nach Vindobona, Carnuntum, Brigetio, Aquincum, das heißt in die Legionslager, und ab 106 in die an der Donau gelegenen Hauptstädte der zweigeteilten Provinz. Hier sind die Stützpunkte der Heereslieferanten beziehungsweise die Kollegien tätig.[12] Ähnlich wie in Carnuntum gab es vermutlich auch im Legionslager von Aquincum ein Serapis-Isis-Heiligtum.

Denn aus **Aquincum** ist von einem Altarstein eine bemerkenswerte Isis Darstellung bekannt. Sie zeigt die Göttin in ihrer "klassischen" Gestalt als Isis-Pharia (Tf. 30/2 und 30/3): am Schiffsbug stehend, in einem von kräftigem Wind zerzausten Gewand, hält sie mit beiden Händen das Segel zusammen. Der hellenistische Prototyp hat sich Jahrhunderte lang weitervererbt ... Die Stilmerkmale des besagten Steindenkmals deuten auf die Wende vom 2. zum 3. Jahrhundert hin. Das Denkmal stammt aus der Retentur des Aquincumer Legionslagers, aus deren Nähe seit Jahrzehnten zwei reliefierte Tafeln bekannt sind; sie mögen

11 Dazu WILD 1984, 1802.

12 FITZ 1978, 71-86.

die Stirnfront des Heiligtums geziert haben.[13]

Aus derselben Periode, Ende des 2. und Anfang des 3. Jahrhunderts, als das Isis-Relief entstand, sind zwei weitere Steindenkmäler aus Aquincum bekannt, die als Serapis-Isis Darstellungen publiziert wurden.

1) Eine kleine Votivtafel aus weißem Marmor (H. 36 cm) mit einem barockartigen Rahmen. In der Bildfläche sitzt "en face" ein Götterpaar, beide tragen den Kalathos auf dem Kopf (Fundort: Aranyhegy, Villengegend der Zivilstadt).[14]

2) Eine viereckige, große Tafel aus Kalkstein (H. 88 cm) mit flachem Rahmen. In der Mitte des Bildes sitzt auf einer Bank das Götterpaar in derselben Haltung wie auf dem erwähnten Votivtäfelchen. Die Göttin, die als Isis identifiziert wurde, hält einen geflochtenen Korb voll Früchte im Schoß. Zur Rechten der Göttin steht eine junge Dienerin in langem Gewand. Der Gott trägt den Kalathos auf dem Kopf, in der Hand hält er einen Schlüssel, vor seinen Füßen sitzt der dreiköpfige Zerberus. Zur Linken der Gottheit ist Merkur im Laufschritt im Profil dargestellt. (Tf. 30/1)

Das Götterpaar wurde in der ersten Publikation als Isis und Serapis betrachtet, später wurden vor allem die chtonischen Züge der Darstellung betont und demnach das Relief als eine Widmung an Pluto-Proserpina gedeutet. In letzter Zeit kam es zur Neubearbeitung des Bildes, wobei der Gott als ein für die Severerzeit typisches Beispiel des Dis Pater-Succellus-Serapis bestimmt wurde (Fondort: Vihar Str. 13, die Retentur des Legionslagers von Aquincum).[15]

Zusammen mit der oben beschriebenen Tafel kam als Pendant des Isis-Serapis Reliefs ein weiteres Steindenkmal zum Vorschein mit den Bildern von Jupiter-Juno. Beide Reliefs charakterisiert derselbe Stil.[16]

"Wie bekannt, war Septimius Severus ein eifriger

13 PÓCZY 1991, 245.

14 KUZSINSZKY 1897, 13.

15 KUZSINSZKY 1897, 13.

16 KUZSINSZKY 1897, 106.

Anhänger des Serapis. Er läßt sich auch gerne mit der Kaiserin in der Gestalt des Serapis und der Isis abbilden. Während den Jahren seiner Regierung erschienen nicht nur in Rom, sondern auch in den westlichen Provinzen Zeus-Serapis Abbildungen, ... wobei sich in dem Ausdruck der Gottheit die Allmacht, ... jedoch mehr die des himmlischen Zeus, als die des unterweltlichen Plutos widerspiegelt." Auf manchen Münzen erscheint das Kaiserpaar zur selben Zeit als Jupiter und Juno.[17]

Auch die Denkmäler einer dritten ägyptischen Gottheit findet man in den beiden Provinzsitzen Pannoniens am Limes: die des Gottes Amun, ursprünglich oberste Gottheit der Stadt Theben. Im Zeitraum der afrikanischen Kaiserdynastie, während der römischen Besetzung, war er der Hauptgott des Hafens von Alexandria und wurde als Schirmherr des Pharos verehrt. (So trat er auch in Ostia auf.)[18] Ab dem Ende des 2. Jahrhunderts stieg er in einen mit Zeus identischen Rang auf, "als Sonnengott...". Seine Heimischwerdung in den Provinzen kann ebenfalls in die Severerzeit gesetzt werden, allerdingst etwas später als die der Gruppe Serapis-Isis. In Carnuntum ist eine Inschrift aus dem Jahre 235 bekannt, die ihn als Jupiter Ammon bezeichnet.[19]

Ein Aquincumer Altarstein entstand wesentlich früher: Marcus Caecilius Rufinus Marianus, tribunus laticlavius, d.h. stellvertretender Befehlshaber, der legio IV. Flavia, hat dieses Votivdenkmal gestiftet, das Hammuni I(ovi) O(ptimo) M(aximo) anspricht. Forscher datieren es in die Jahre vor 180. Die Familie stammte aus Afrika, aus Theveste.[20]

Vielleicht noch wichtiger als der obige Altar sind die Ammondarstellungen aus Aquincum, besonders interessant sind die Stellen, an denen sie angebracht waren:

a) Eine Ammondarstellung ziert das korinthische Kapitell der

17 VIDMAN 1970, 149 und Anm. 48.

18 JOBST 1992, 126; MÓCSY 1990, 258.

19 JOBST 1992, 75.

20 CIL III, 3463; ALFÖLDY 1963, 53, Anm. 66, Abb. 8.

Säulen am Haupteingang des Kaiserkult-Tempels (Seve-rerzeit)[21], der aus dem am Forum der Zivilstadt stehenden ehemaligen kapitolinischen Heiligtum umgestaltet worden war. (Tf.31/1)

b) Die Konsole (Tf. 31/2-5) über einem der Eingänge zum öffentlichen Bad des Legionslagers dekorierte man mit dem Kopf des Zeus-Amun (zweite Hälfte 3. Jahrhunderts).[22] Er wurde in der Gestalt des Jupiter-Ammon beziehungsweise Ammon-Sol als triumphaler Herrscher verehrt. Ab dem Ende des 3. Jahrhunderts, als zuerst Kaiser Aurelianus (270-275) und dann Kaiser Diocletianus (294-305) den Kult auf die Ebene der Staatsreligion heben, setzt sich der Charakter des Sol durch. Seines ägypti-schen Ursprungs und seiner Wurzeln beraubt, dient er in allen Provinzen der Einheit des Imperiums.[23]

Damit sind wir in einer Zeit angelangt, in der sich bemerkenswert erweise die engste Beziehung zwischen Pannonien und Ägypten zeigt, zumindest im Hinblick auf die Importe beziehungsweise Handelsartikel. Neben den aus westlichen Provinzen eintreffenden Versorgungsgütern des vorangehenden Zeitalters erhalten die Heereslieferanten auch für Waren aus Ägypten und Nordafrika Vergünstigungen.

Am Ende des 3. und in den ersten Jahrzehnten des 4. Jahrhunderts hat man in Pannonien mit einem weiteren Kult ägyptischer Herkunft zu rechnen, worauf die Mumienbestattungen unser Augenmerk lenken. Bisher gibt es lediglich aus dem Bereich der größeren Legionslager entlang des Limes Angaben über Freilegungen: aus der Militärstadt, d.h. der *canabae legionis*, von Carnuntum, Brigetio und Aquincum beziehungsweise vom Standort eines der syrischen Bogenschützenregimenter in Intercisa.[24] Das Mumifizierungsverfahren der Spätkaiserzeit ist bereits oberflächlicher als das im frühen Reich angewandte, aber identisch mit dem gleichzeitigen ägyptischen des 2.-3. Jahrhun-

21 NAGY 1935, 25.

22 KABA 1991, 56, Abb. 93-94.

23 FITZ 1998, 126.

24 PÓCZY 1997-1998, 59-76.

derts und auch mit den aus der Stadt Rom bekannten, für die gleichen Periode typischen Mumien der Katakomben.[25]

Mumifizierungen durften ausschließlich von Angehörigen der Priesterkaste vorgenommen werden, das Verfahren galt als Geheimniss und sehr kostspielig. Bemerkenswert ist, dass man unter den Grabbeigaben der Bestatteten auch zahlreiche ägyptische Importe findet. Überhaupt ist dies das Zeitalter, in dem die über Italien und aus dem Osten auf dem Wasserweg der unteren Donau eintreffenden Importe auf ägyptische Beziehungen hindeuten. In den vorangehenden Jahrhunderten zählte so etwas als Seltenheit. Geliefert wurden Luxusartikel, feine Keramik, Terra sigillata chiara, besondere Gläser, gemusterte Stoffe und mit Seide gewebte Textilien.[26]

An dieser Stelle möchte ich anmerken, weil es sich um eine auffallende zeitliche Übereinstimmung handelt, dass mit der Verbreitung des Arianismus in Pannonien der Brauch der Mumifizierung unverhofft abbricht. Anscheinend vermochte es diese religiöse Gruppe Mitte des 4. Jahrhunderts nicht, den Kampf mit den Vertretern der neuen, kämpferischen Geistesströmung aufzunehmen.[27]

Dem skizzierten Ablauf zufolge fanden die ägyptischen Kulte in Pannonien stets in den Städten Verbreitung, ihrer religiösen Gemeinschaft gehörte die Führungsschicht der städtischen Einwohnerschaft an. Beachtung verdient, dass sich im Zeitraum von 1. bis zum 4. Jahrhundert n. Chr. der Schwerpunkt innerhalb der Strömung viermal verlagert, wobei immer eine jeweils andere Gottheit in den Vordergrund gelangt. Isis, Serapis und Isis, Ammon und nach ihm der Brauch der Mumienbestattungen repräsentieren die unterschiedliche Geisteshaltung der Religionen.

Die Quellen weisen darauf hin, dass die immer neuen

25 NAGY 1935, 11-12 und Anm. 3-4; E. SWOBODA 1964, 199 und 294.

26 GABLER 1990, 201 und 203.

27 KRAUTHEIMER, 1987, 30-31, 33, 39.

religiösen Tendenzen an konkrete ökonomische Prozesse zu binden sind: Die Verehrung von Isis, in den Städten an der Bernsteinstraße, die den Waren der mediterrane Gewässer befahrenden Handelsvölker auch auf dem Festland Schutz gewährte, hing eng mit den direkten norditalienischen Handelskontakten an der Wende des 1./2. Jahrhunderts zusammen. Savaria, die Hauptstadt des ungeteilten Pannonien, ist in diesem Zeitraum der Einrichtung der neuen Provinz das natürliche Zentrum des Isis-Kultes.

Während einer folgenden Periode fällt unter den beiden Gottheiten des ägyptischen Götterpaares dem als "Heiland" verehrten Serapis eine wichtige Rolle zu. Die aus Carnuntum und Aquincum bekannten Serapis-Isis-Heiligtümer und Inschriften entstammen einem Zeitraum, als die Heeresversorgung in Pannonien nicht mehr aus Italien, sondern entlang des Rhein-Donau-Limes erfolgt. Der Kult der besagten Gottheiten in den beiden Provinzhauptstädten des geteilten Pannonien unterstreicht diesen Vorgang.

Als dritte Stufe kann die Rolle Ammons betrachtet werden – ebenfalls im Bereich der Legionslager und hauptsächlich in der zweiten Hälfte des 3. Jahrhunderts, als die Verbindung mit dem ägyptischen Heer sehr eng ist.

Das Vorhandensein der Mumiengräber schließlich verrät deutlich, dass eine dünne Führungsschicht der Gesellschaft die neue Kultgemeinde bildet. Mitte des 4. Jahrhunderts verliert sie unvermittelt ihre Positionen, als der über längere Zeit in Pannonien weilende beziehungsweise von hier aus regierende Kaiser Constantius II. völlig unter den Einfluss der Arianer gerät.[28]

Wie allgemein bekannt, wurde der aus Ägypten stammende Isis-Kult im Rom des 4. Jahrhunderts zum Symbol des Widerstandes gegen die christlichen Bestrebungen Kaiser Konstantins I.[29] (Theodosius I. erklärte das Christentum 392-394

28 ALFÖLDI 1942, 708, 713.

29 PÓCZY 1997-1998, 65.

zur Staatsreligion.) Erst im Jahre 535 verbot Justinianus I. den Isis-Kult. Doch im Verlaufe der etwa anderthalb Jahrhun-derte dauernden Machtkämpfe haben sich die Ausdrucksformen der althergebrachten Isis-Religion und des neu aufgekommenen Christentums derart aneinander angeglichen, dass der Unterschied in ihrem archäologischen Nachlass in vielen Fällen kaum erkennbar ist. Auch die Riten und Symbole der gegnerischen Seiten – auf der einen die Mithras-Gläubigen und Isis-Anhänger, auf der anderen die Christen – unterscheiden sich kaum. So mag aus der Darstellung der Isis, die ihr Kind Harpocrates im Arm wiegt, bis zum 5. Jahrhundert das Bild der Jungfrau Maria mit dem Gottes-sohn im Schoß geworden sein...[30]

Klára Póczy
Aquincum Museum
Budapest

30 JOBST 1992, 75.

Topographische und chronologische Problemen des Iseums von Savaria

Zusammenfassung

Das aller wichtigste pannonische Momument für die Kulte mit ägyptischem Ursprung in Ungarn ist das Iseum in Szombathely. Die Ergebnisse der Ausgrabungen wurden letztes Mal von István Tóth im Jahrbuch Savaria zusammengefasst, das am Beginn dieses Jahres erschien. Er ist auch auf die Zweifel und Probleme eingegangen, die im Laufe der letzten 45 Jahre von der heimischen und internationalen Forschern verfasst wurden. (Die Ergebnisse und Probleme der Forschung des Iseum savariensis, Savaria 22/23, Szombathely, 1997-1998)

In meinem Vortrag möchte ich mich zu den Gedanken und Fragen, die in dem Artikel von István Tóth formuliert wurden, und die im Titel angeführt sind, äussern, und die neuesten und wichtigsten Ereignisse der Ausgrabungen in dem Territorium des Iseums und der Umgebung des Bezirks des Heiligtums zusammenfassen.

1. Charakter und zeitliche Zuordnung der Bebauung der Umgebung des Iseums: Die Ausgrabungen der letzten Jahre haben die relativ frühe Bebauung dieses Territoriums, die mit dem letzten Drittel des 1. Jahrhunderts begann, bestätigt. Die ständige, dichte und anspruchsvolle Bebauung regte wieder die frühe Datierung der Bebauung des Iseums an, bzw. die Existenz eines früheren – vielleicht mit der Holzperiode der Nachbarbereiche zeitgleiche – Heiligtums im 1. Jh. n. Chr. Aufgrund der Kenntniss der relativ genauen Stratigraphie des Bereiches, der Podiumtempel des Iseums, die grosse Säulenvorhalle, bzw. die Niveaudifferenz zwischen den Tabernae, die den Hof beiderseitig begrenzen,

kommen Zweifel gegen die jetzige Periodengliederung des Heiligtums auf.[1]

2. Besonderheiten des Grundrisses des Temenos des Iseums: Analogien des charakteristischen Grundrisses des heiligen Bezirkes sind sowohl in anderen orientalischen Kultbereichen, als auch in anderen sakralen Gebuden mit profaner Frbung zu finden. Von der jetzigen Rekonstruktion fehlen charakteristische Gebude (wie zB. *lavacrum-purgatorium*), die in einigen anderen Isis Heiligtümer vorkommen, und die auch in diesem Temenos vorstellbar sind, worauf die Holzreste, die im Laufe der Ausgrabungen gefunden wurden, hindeuten können. Von den möglichen Rekonstruktionen des heiligen Bezirkes an der heute noch nicht freigelegte westlichen Seite können wir schon mit grosser Wahrscheinlichkeit die spiegelsymmetrische westliche Abschlieung – ähnlich wie an der östlichen Eingangshalle – in der Kenntniss der Topographie des Territoriums ablehnen.

Ottó Sosztarits
Savaria Museum
Szombathely

Korrekturzusatz:
Auf Grund die Ergebnisse der neuen Ausgrabungen des Iseums, die der Autor mit Tihamér Szentléleky, Zsolt Mráv und Tamás Mezõs im Sommer 2001 begonnen hat, kann Punkt 2. vollkommen überwunden betrachtet sein. Siehe: O. Sosztarits (Hrg), A savariai Izeum – The Iseum of Savaria, Szombathely 2002, im Druck.

1 Neustens: B. Balogh-O. Sosztarits-A.B. Szilasi, A szombathelyi Éva malom területén feltárt korarómai falfestmény és topográfiai jelentõsége. (Das auf der Gelände der Éva-Mühle freigelegte frührömische Wandgemälde und dessen topograhische Bedeutung), in: Szentléleky Festschrift. Savaria 25/3, 2000-2001, 155-190.

ELÉMENTS D'INTERPRÉTATION DES AUTELS DE SPHINX DE PANNONIE[1]

I. Nous connaissons de Savaria et de ses environs deux autels en grès de dimensions moyennes, dénués d'ornement (Figure 1), qui ont été dédiés par leurs constructeurs de jadis, d'une manière peu fréquente[2], soit aux Sphinx majestueux, soit au Sphinx, ou d'une façon complémentaire aux Sphinx aussi. De même nous connaissons d'Aquincum un fragment d'inscription qui contient lui aussi une dédicace adressée aux Sphinx. (Figure 1a) De la dédicace du fragment d'autel RIU 303, il subsiste le mot Sphinx, que nous pourrions compléter à la forme de Sphinx[i], bien que par manque de place même la présence de la lettre I soit douteuse. Nous pouvons de même supposer qu'il s'agit de l'écriture abrégée d'un nom, et dans ce cas, en connaissance de la dédicace de l'autre autel, celle-ci serait probablement Sphinx[i(bus)] ou Sphinx[(ibus)], suivi du groupe de caractères *METILI* dont le sens est difficile à cerner de plus près. Ce dernier pourrait désigner une partie du nom du constructeur que nous pouvons compléter à la forme de Metili[us] ou Metili[a], dont Metili[a] est plus vraisemblable. (Figure 1b) Sur le fragment d'autel RIU 104[4] le texte restant est un peu plus long. La dédicace

1 Nous pouvons lire en allemand à peu près le même raisonnement dans Folia Archaeologica 2002.

2 „D'une autre provenance, nous ne connaissons pas de dédicaces semblables" – TÓTH 1998a, 9. cf. aussi MALAISE 1984, 1615-1691, où il ne mentionne pas les sphinx de la région traitée.

3 RIU = (CIL), III, 10913 = RSS, 43 et BUOCZ 1964, 274-275.

4 L'autel est déposé dans le département lapidaire du musée de Szombat-

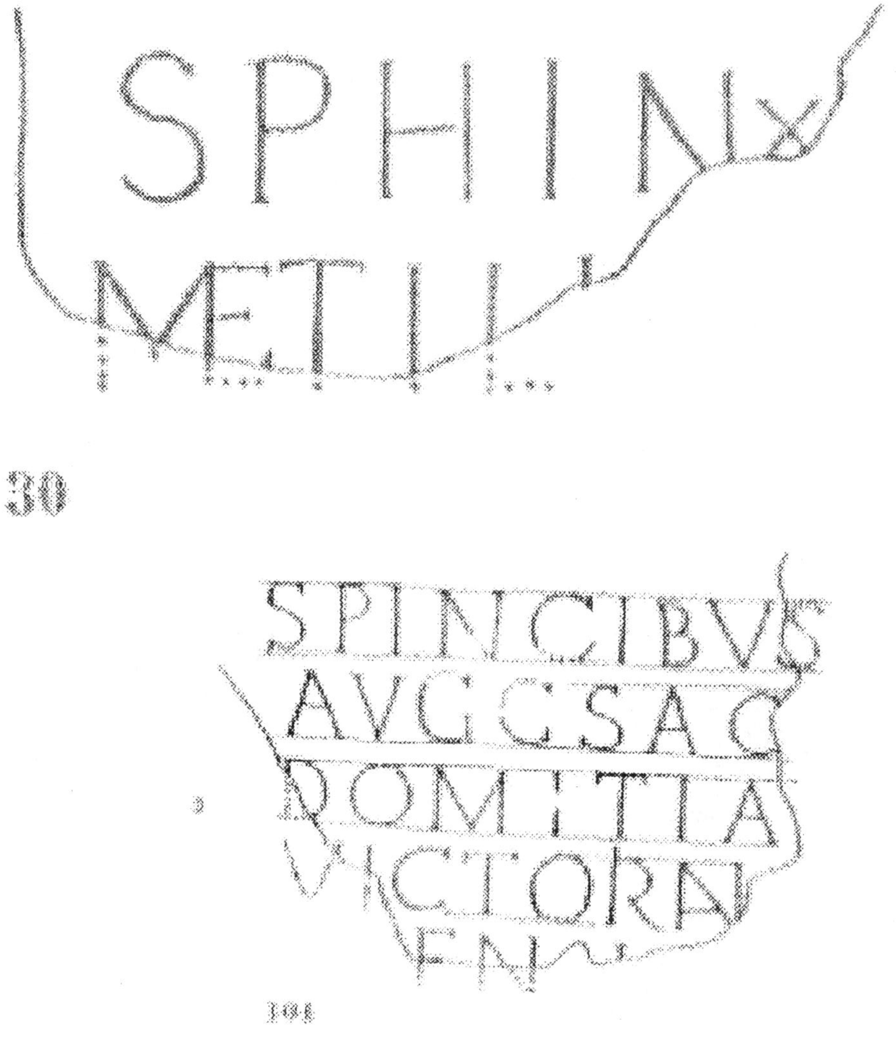

Fig. 1. Les desseins des inscriptions des autels dédiés aux Sphinx à Savaria après (a) RIU 30, et (b) RIU 104. La provenance de l'autel (a) est certaine: terrain en face de l'Iseum. L'autel (b) se trouve à Szombathely, dans le département lapidaire du Musée Savaria. La provenance exacte n'est pas connue. Il fut trouvé dans la proximité de l'Iseum.

6460 ara. Alt-Ofen in aedibus n. 726 Antonii Flesser.

//SPHINC/////
///·INAEI////
//// ≥D7^ ///
//////////

Rómer *Archaeol. Értesitö* 1 (1869) p. 60.

Fig. 2. L'inscription fragmentaire de l'autel trouvè à Aquincum. Après CIL III.

que nous pouvons y lire sans ambiguïté est Sp(h)incibus / Augg(ustis) sac(rum). Conformément à l'autel précédent, à cause de son caractère fragmentaire, le nom du dédicant ne peut être lu sans ambiguïté, bien qu'il en reste plus. Il peut être lu comme Domitia Victorina (ligatures R+I et N+A dans le cognomen) désignant un constructeur féminin, de même que Domitia(nus) Victorian(us) désignant un constructeur masculin. De ces deux variantes, la première, c'est-à-dire l'interprétation en tant qu'un nom féminin me semble plus probable, d'un point de vue épigraphique aussi. La dernière ligne de ce qui reste de l'épigraphe forme un groupe de caractères incompréhensible et fragmentaire. Les caractères juxtaposés FN ou EN peuvent désigner éventuellement la dernière lettre d'un mot abrégé et le

hely. Son lieu d'origine précis nous est inconnu – voir aussi CIL, 3, 10914 = RSS, 42 et BUOCZ 1964, 275.

début du mot suivant, auquel appartient probablement une lettre I, situé à un espace de caractère plus loin. (Figure 2) La partie subsistante de l'inscription d'Aquincum (CIL, III, 6460)[5] est la suivante, d'après le CIL : [—-]Sphinc[—-] / [—-]•INAEI[—-] / [—-]ZD + ʌ [—-] / [—-]. Sa lecture ne peut être complétée avec certitude qu'à l'égard de la dédicace : Sphinc[ibus]. Au début de la ligne, la partie cassée, connue de sa description, ne désigne vraisemblablement que l'absence du bord, du cadre du champ d'écriture. Avec le mot Sphinc[ibus], la dédicace peut être considérée comme la première ligne dans son intégrité. Dans la seconde ligne se trouvait le nom du constructeur de l'autel, que nous ne pouvons plus compléter aujourd'hui. La liaison des voyelles donne des consonances grecques au mot fragmentaire, par conséquent il nous fait penser plutôt à une personne féminine d'origine orientale qu'une forme au génitif pouvant désigner une filiation. Cette dernière possibilité est assez douteuse, vu le peu de place qui reste devant le fragment de mot, insuffisant pour inscrire un nom entier. L'interprétation des fragments de caractères de la troisième ligne nous est encore impossible.

Les différentes inscriptions de noms de dieu, que nous pouvons observer dans la dédicace des autels, peuvent être dues aux différents usages de ces mots – dont plusieurs exemples sont connus concernant le sphinx[6] –, mais il peut s'agir également de leur écriture erronée. Quant à la dédicace, le second autel nous paraît plus intéressant. Le mot Sp(h)incibus pourrait désigner un groupe de divinités, de nombre incertain, mais la lettre G doublée à cause de la forme abrégée de l'élément suivant du nom de dieu – Augg(ustis) – se réfère plutôt à deux divinités distinctes[7]. Les trois monuments en pierre sont certainement des autels, à savoir des autels votifs, érigés d'un geste religieux. Quant à la date de

[5] RÓMER 1869, 60. – Alt-Ofen im Garten des Hauses no. 726 – il est probablement perdu depuis.

[6] Cf. HERBIG 1929, 6. 3, Neue bearb. 1703-1749, 1703-1704 ; LTL, 448.

[7] Voir DEMISCH 1977, 106 – Sphingenpaare, 231 – Die ägyptische Doppelsphinx.

leur érection, les IIe et IIIe siècles après J.-C. nous paraissent vraisemblables[8].

Jusqu'à présent, personne ne s'est occupé des autels d'Aquincum, tandis que de nombreux chercheurs se sont intéressés à ceux de Savaria. En conséquence, ces derniers ont suscité plusieurs interprétations, peu détaillées à cause du caractère irrégulier du phénomène. Les conclusions admissibles, de même qu'inadmissibles, peuvent être adaptées à l'autel d'Aquincum aussi. D'après Lajos Balla, à Savaria le culte des sphinx est le résultat de la fusion des conceptions religieuses orientales et locales, à savoir celtes[9]. Mais ailleurs, il écrit que l'on trouve des statues, des reliefs et des épigraphes (? nous ne connaissons pas d'autres épigraphes) dans des sanctuaires égyptiens d'Isis et dans le temple d'Isis du champ de Mars de la ville de Rome, tous dédiés au(x) sphinx[10]. A part l'autel de sphinx d'Aquincum, je n'ai pas retrouvé d'inscriptions semblables[11]. Selon l'opinion de Vilmos Wessetzky, les autels de sphinx de Savaria appartiendraient aux vestiges du culte religieux en Haute-Pannonie des divinités égyptiennes[12]. Selon une opinion divergente, les *sp(h)inces* représenteraient les figures d'un monde divin particulier qui nous ne pouvons pas décrire avec l'*interpretatio romana* de Tacite[13]. Cette opinion peut être pleinement soutenue, tout en précisant que d'une part, d'après ses représentations, en Italie on connaissait depuis longtemps un être semblable désigné aujourd'hui du nom de sphinx, et que d'autre part, ce phénomène particulier ne peut être substitué

[8] BUOCZ 1964, loc. cit.

[9] BALLA 1971a, 474.

[10] BALLA 1971b, 37-45, 42. Sur les différentes statues de sphinx des temples d'Isis de la ville de Rome voir par exemple : ROULLET 1972, 23-51, Cat. No. 277-313.

[11] CIL, III, 6460.

[12] WESSETZKY 1959, 265-282, p. 271 et WESSETZKY 1961, 36.

[13] TÓTH. 1998a, 13.

par rien d'autre. Par conséquent, l'interprétation à la romaine[14] de ses aspects provenant éventuellement d'autres régions est impossible, ainsi il nous reste seulement de les évoquer à leur propre nom, tout comme d'autres divinités orientales. Les diverses opinions se rapprochent sur le point qu'il s'agit essentiellement de(s) divinité(s) apotropaïque(s)[15].

Quant aux antécédents du culte du sphinx, parmi les peuples de la partie orientale de la Méditerranée le sphinx était un phénomène habituel, voir populaire. En connaissance des trois épigraphes pannoniennes de l'époque impériale et dans le cadre de cet exposé, je n'ai pas la possibilité d'entrer des les détails de l'arrière-plan culturel et religieux de ce phénomène[16].

Dans l'ordre chronologique, c'est en Egypte que nous rencontrons le sphinx pour la première fois, où il était à la fois le défenseur et l'incarnation du souverain[17]. Nous le retrouvons dans la culture minoenne et mycénienne, de même qu'en Préasie, en Syrie et en Phénicie, nous le voyons en Grèce et en Italie, et finalement dans les provinces romaines conquises après Jésus-Christ. Autrement dit, les représentations de la figure du monstre fabuleux et divin peuvent être suivies dans les temps à travers l'histoire jusqu'à l'avancée définitive de la chrétienté[18]. Sous quelle forme apparaît-il, le sphinx? Dans le

[14] Pour l'interprétation des figures divines égyptiennes cf. aussi VIDMAN 1970, 106-124 ; MALAISE 1984, 1624.

[15] BALLA 1971b, WESSETZKY 1959, loc. cit. ; cf. aussi RENARD 1968, 273-305. Pour un aperçu général voir DEMISCH 1977 – avec une littérature supplémentaire ; voir aussi DESSENE 1957 – avec une littérature supplémentaire.

[16] DEMISCH 1977.

[17] ILBERG 1909-1915, 1298-1508, et 1301-1302. Pour un aperçu général du sujet avec une vaste matière d'illustration de l'Egypte à Rome, de même que KRAUSKOPF 1994, 1-15 et une riche illustration loc. cit. VII/2, 6-15 ; de même que DESSENE 1957; DEMISCH 1977.

[18] Surtout concernant l'époque classique voir HERBIG 1929, 1703-1749 et ILBERG 1909-1915 ; cf. aussi NICOLE 1908, 1431-1439 ; cf. aussi DEMISCH 1977, 101 et DESSENE 1957.

monde antique, sa figure se retrouve fréquemment sur une multitude de monuments plastiques de petite et de moyenne taille, des représentations peintes, des objets d'usage, souvent en tant que simple élément de décoration, mais nous le voyons également sur des autels tombaux, des sarcophages, représenté seul ou par deux[19]. D'après Suétone, au début de l'Empire, il apparut sur le sceau d'Auguste[20], sous Hadrien sa figure à quatre corps décorait le revers des médailles[21]. A quoi ressemblait-il exactement ? D'après de nombreuses descriptions et illustrations, sa figuration était relativement changeante, ce qui ne facilite pas la définition de sa fonction générale. On le représentait surtout comme un être ailé, mais des fois sans ailes aussi. Une représentation relativement simple est la figuration avec la tête d'un homme et le corps d'une lionne, comme nous le connaissons de l'Egypte, où sa figure sans ailes apparut pour la première fois. Une représentation plus complexe est celle de l'être ayant le visage et les bras d'une jeune fille, les doigts d'un lion, la parole d'un homme, le corps d'un chien, la queue d'un dragon et les ailes d'un oiseau, qui apparaît ailleurs simplement avec un visage de jeune fille, un corps de lionne et avec des ailes. En Orient antique, on le représentait debout aussi, tandis qu'en Egypte en mouvement ou en se battant. Quant au monde gréco-romain, sa figuration assise ou couchée est beaucoup plus fréquente. Sur ses représentations pannoniennes, dont deux se trouvent dans le département lapidaire du Musée national de la Hongrie, apparaît une scène bien connue de la mythologie grecque, la rencontre d'Œdipe et du Sphinx, cette fois taillée en pierre[22]. Selon le mythe, le Sphinx fut envoyé à Thèbes soit par

19 ILBERG 1909-1915 ; KRAUSKOPF 1994, 6-15 ; DEMISCH 1977, 106.

20 Suetonius, Augustus 50. – le sphinx représenté dans la pierre de son anneau sigillaire est certainement le symbole de la souveraineté.

21 DEMISCH 1977, Abb. 81; NOLL 1955, 67-74, et 72-73 Abb. 43.

22 MNM 62. 86. 1 - Lapidaire no. 60, lieu de découvert inconnu, du IIe siècle après J.-C. ; MNM 62. 87. 1 - Lapidaire n. 59, Óbuda, du IIe siècle après J.-C. Tous deux faisaient partie d'un monument funéraire.

Héra, soit par Dionysos, ou encore par Hadès, pour punir les habitants de la ville. Au mont Phinikion ou sur l'agora de la ville, une vierge de lionne, appelée aussi l'étrangleuse, proposait des énigmes aux habitants, et si ces derniers n'étaient pas capables de les résoudre, elle choisit l'un d'entre eux et le tua. L'Œdipe au sort malheureux trouva la solution de l'énigme, sur quoi le Sphinx se précipita du haut de son rocher et périt. Selon une autre version, c'est Œdipe qui l'abattit d'un coup de lance. Par sa destruction il sauva la ville, dont il devint le roi[23]. La majorité des représentations de sphinx de l'Antiquité puisait de cette histoire[24]. De plus tard, nous n'avons plus de données appréciables sur les sphinx du monde oriental[25] dans les provinces de l'Empire romain. Le monde qu'ils représentaient donna sa place à un autre, que l'Egypte et la mythologie grecque nourrissaient à l'égard des sphinx. En Egypte et dans la vie religieuse à base égyptienne, le sphinx gardait son rôle précédent. A l'époque romaine, il était dénommé « Sphinx », indépendamment de son sexe, c'est-à-dire que l'on représentait avec une tête d'homme ou de femme, de même que de sa forme d'apparition. D'un point de vue iconographique, les sphinx de l'Empire romain suivent souvent le sphinx d'Œdipe, bien connu des représentations hellénistiques, quoi qu'il en soit le contexte ou l'arrière-plan idéologique[26]. Quant à son sexe, d'après son nom féminin le sphinx est du sexe féminin. Le Sphinx d'Œdipe était en effet un être féminin : une « vierge de lionne ». Le nom de « sphinx » désignait également les êtres sculptés avec un corps de lion et une tête humaine. Ils appartenaient cependant – à part quelques exceptions – au sexe masculin. Ils s'agissaient des statues constituées par le corps d'un lion et la tête d'un pharaon, des

23 KERÉNYI 1977, II. 1. X p. 248-250.

24 Cf. KRAUSKOPF 1994.

25 Pour un aperçu bref sur les sphinx du Proche-Orient voir *Sphinx*, HAUSSIG 1965, 535-536 – le rôle qu'il jouait dans la région n'est pas tout à fait claire ; voir aussi DEMISCH 1977, 40.

26 I. KRAUSKOPF 1994,

représentations où le sexe ne jouait qu'un rôle secondaire selon celui du souverain actuel. L'essentiel de la représentation consistait en effet dans la figuration du pharaon en tant que souverain.

En somme, dans le monde hellénistique, et surtout des les provinces romaines des IIe et IIIe siècles après J.-C., lorsque l'on parlait de sphinx – qu'il s'agissait d'un sphinx de pharaon ou dieu solaire égyptien, du sphinx d'Œdipe ou de la représentation visuelle de l'être que l'on nommait sphinx, on pensait en général à l'image du sphinx ailé d'Œdipe, ayant un corps de lion et un visage de jeune fille, donc à une image compréhensible pour tous. Vu la désignation communément utilisée et la multiplicité des ces représentations, le sexe du sphinx n'aurait pas eu beaucoup d'importance, ce qui est certes encore plus valable pour l'époque romaine. Ce qui importait, c'était la signification que l'être représenté portait dans son caractère de sphinx.

En ce qui concerne l'interprétation des sphinx des autels de Pannonie, deux aspects nous paraissent déterminants : 1. ils apparaissent dans les dédicaces d'autels, 2. ils ont été retrouvés à Savaria ou dans ses alentours, c'est-à-dire dans la ville qui constituait le plus grand centre religieux de la Pannonie, et de plus – comme le sanctuaire d'Isis-Séraphis[27] le prouve – un lieu de culte extrêmement important des dieux d'Egypte et d'Alexandrie, avec une vaste communauté de fidèles. C'est de ce point que nous pouvons mieux saisir l'essentiel des sphinx divins de Savaria, et c'est ce qui peut être appliqué pour l'autel d'Aquincum aussi.

Le premier point de vue – qui ne nécessite pas beaucoup d'explications – est le suivant : ceux pour qui on élevait des autels dans l'Empire étaient des dieux. Cela exclut évidemment du cercle des sphinx-dieux le sphinx grec, étant une créature mortelle, détruite par l'instrument des dieux, un héros mortel, qui par conséquent n'était pas d'origine divine. Ainsi le sphinx du

[27] MRÁV 2000, p. 67-97.

mythe d'Œdipe en tant que support de l'essence divine – mise à part sa figure très variée[28] – n'a pas de rapport étroit avec les figures divines, vénérées d'autel dans le centre pannonien du culte des divinités égyptiennes et alexandrines que nous traitons. Si nous le retrouvons sur les monuments funéraires, c'est pour la même raison qu'y apparaît la tête de la gorgone Méduse, c'est-à-dire grâce à son caractère apotropaïque. Toutefois, sa présence sur les monuments funéraires entre organiquement dans l'ensemble des unités de construction en pierre des tombes, ornées de sculpture à sujet mythologique gréco-romain.

Etant donné qu'à l'époque le culte de la figure préromaine du sphinx n'entre plus vraiment ou plus du tout en jeu, il ne reste qu'un milieu qui nous permet d'approcher de l'interprétation des dédicaces pannoniennes de Sp(h)incibus, Sphinc[ibus] et Sphinx[ibus], à savoir l'Egypte, ce qui correspond à notre deuxième point de vue[29].

Sur un relief datant de l'Empire de la collection de Kunsthisorisches Museum à Vienne, nous retrouvons le sphinx portant sur sa tête des symboles égyptiens : un disque solaire et des cornes de bélier. Nous n'en avons pas de datation plus exacte à cause du manque d'études relatives, comme Rudolf Noll l'a souligné en 1955[30]. Le relief représente de face et en mouvement le sphinx à tête d'homme et à corps de lion, bien connu de l'Egypte. Le sphinx à queue de serpent et à corps de lion a le visage d'une jeune fille. Au-dessus de lui nous voyons assis le griffon ailé, connu du monde hellénistique[31]. La déesse se tenant debout à côté du sphinx est apparemment Athénée ou

[28] Voir DESSENE 1957, par exemple XIX 258 et XX 262, 269 ; cf. KRAUSKOPF 1994, texte et illustration.

[29] Cf. WESSETZKY 1961, 36.

[30] NOLL 1955, 67 cite les lignes d'A. Scharff, qui déjà en 1939 a attiré l'attention sur le manque d'études y afférentes, ce qui rend difficile toute datation relative à l'époque romaine.

[27] MRÁV 2000, p. 67-97.

Némésis[32]. Ce relief est un élément qui justifie que nous poursuivons nos recherches sur le sphinx de Savaria soit en Egypte, soit ailleurs, mais toujours dans un milieu en rapport avec l'Egypte. En récapitulant brièvement : en Egypte le sphinx est le symbole du souverain et du soleil, de même que le défenseur de ce premier[33]. Les sphinx égyptiens à pied et corps de lion furent également les symboles du Nil montant, dont les crues arrivaient aux mois de juillet et d'août, sous les constellations du Lion et de la Vierge[34]. Sa figure ne peut être confondue avec rien d'autre, et nous pouvons affirmer que ce n'est qu'en Egypte ou sous les influences égyptiennes que les êtres fantastiques du monde antique, appelés sphinx remplissaient une telle position qui permettait que même à l'époque romaine on dédie des autels à l'en-soi qu'ils représentaient[35]

II. Au début du XIXe siècle, la collection d'objets d'art du comte József Kemény contenait une statuette de sphinx en bronze (Figure 3)[36], dont une épine d'acier sortait de la gaine, servant à sa suspension, probablement au bout d'une barre ou porte-emblème[37]. Une inscription formée de 19 lettres encerclait les trois côtés de la gaine. L'objet retrouvé à Torda, c'est-à-dire à Potaissa romain a été publié pour la première fois en 1847[38]. Au milieu du désarroi de la révolution et de la guerre d'indépendance hongroise de 1848-49, le Sphinx de Potaissa disparut. Comme en 1847 plusieurs personnes ont décrit, dépeint et

32 NOLL 1955, 68 Abb. 40. et p. 69.

33 ILBERG 1909-1915, 1301-1303 ; cf. aussi NOLL 1955, notes 69 et 2.

34 LTL, 448 (J. Facciolati-Aeg. Forcellini - J. Furlanetti).

35 ILBERG 1909-1915, 1303-1304.

36 VLASSA 1980, 133-153 – avec l'histoire et la littérature intégrale des recherches précédentes. D'après VLASSA1980, 152 et M. Bărbulescu, la statuette aurait pu être amenée directement de l'Egypte. BĂRBULESCU 1994, 165-166 – les immigrés orientaux – égyptiens au IIIe siècle après J.-C.

37 La statuette d'Isis de Pollentia pouvait remplir la même fonction, Voir BELLIDO 1967, 118.

38 Pour l'histoire détaillée de la recherche voir VLASSA 1980, 133-136.

Fig. 3. La seule image subsistée représentant la sphinx-statuette de Potaissa, publiée pour la première fois en 1847 dans le no. 188 d'Illustrirte (sic) Zeitung à Leipzig. Le texte écrit en miroir dans son ordre original figure devant l'épine d'acier. Plus bas, le texte est dans sa forme positive, comme il pouvait être incisé dans le moule.

étudié l'objet, la gravure réussite qui en a été faite ne laisse aucun doute de l'existence de son modèle. Plusieurs personnes ont tenté de déchiffrer l'inscription, mais seul Nicolae Vlassa l'a réussi d'une manière convaincante[39]. En ce qui concerne le lieu d'origine de la statuette portant l'inscription, il est à noter que de la province de Dacie, c'est n'est que de Potaissa que nous connaissons un collegium au service du culte d'Isis[40], avec un grand-prêtre à titre de pater, nommé C. Iulius Martialis, et un fonctionnaire à titre de quaestor, nommé L. Livius Victorius, en plus de nombreux autres fidèles[41]. Le collegium Isidis, la personne du grand prêtre et du fonctionnaire, de même que d'autres données portent à croire qu'il existait un temple d'Isis à Potaissa, qui fut en Dacie l'un des plus importants centres religieux où l'on vénérait des divinités égyptiennes[42]. En se référant à l'opinion de Lajos Balla, selon laquelle le sphinx apparaît dans certains sanctuaires d'Isis[43], le sphinx de Potaissa disposait donc d'un milieu naturel où on l'utilisait d'une manière logique et intentionnelle[44], tout comme les autels de Savaria qui furent érigés dans leur propre milieu naturel[45]. Sans la possibilité d'une désignation de lieu

39 VLASSA 1980, 137-142.

40 CIL, III, 882 - Isidi Myrioni/mae C. Iul. Mar/tialis pater et / L. Livius Victori/nus quaestor / collegio Isidis / d. d.

41 CIL, III, 881 - Isidi / et Serapi / C. Iul. Ant/igonus 7 / leg. V. M. P. [C] / et Fl. Apol/linaria / eius v. l. s.

42 Statuettes en bronze d'Harpocrates, d'Ammon, d'Anubis et de Séraphis, relief d'Apis, médaille de Sol-Séraphis, cf. M. BĂRBULESCU 1984, 1677-1680.

43 BALLA 1971, loc. cit.

44 BĂRBULESCU 1994, 164-166 ; BĂRBULESCU 1984, 30-31, 113, 136 ; POPA 1983, 71-79, et 72-75 ; MALAISE 1984, 1677-1680.

45 Cf. BALLA 1974, 1-12 ; BALLA 1971 ; WESSETZKY 1959, TÓTH 1998 ; SYENTLÉLEKY 1960 ; TÓTH 1977, 131-148 ; TÓTH 1996, 1-17 ; TÓTH 1998, 329-352, MRÁV 2000, etc. – les œuvres référées avec toute la littérature précédente. Pour un résumé du culte pannonien des dieux égyptiens voir WESSETZKY 1961.

plus exacte, seul d'après les traces de l'existence des cultes d'origine égyptienne, nous pouvons constater la même chose de l'autel d'Aquincum aussi[46].

Mais revenons au texte. Selon son déchiffreur, le texte est écrit en miroir sur la gaine, étant donné que dans le moule l'inscription fut marquée dans sa forme positive[47]. Cette technique de moulage explique de même le grand nombre de lectures différentes suivant les essais de lecture de gauche vers la droite. Sur les traces de cette découverte, les caractères de l'épigraphe prennent la forme des mots signifiants. Il s'agit des acclamations ou noms de dieux suivants : Ιαω (=Ιαω) Ιωα (= Ιωα) Τιθιε Ρε-Αρμαη[48] – Iao, Iao Tithie(s) et Re-Harmae (=Harmachis)[49]. Tithie(s)-Tithoes qui figure en troisième lieu sur la liste prépharaonique de Manethon, était un dieu solaire égyptien, représenté sous forme de sphinx. Rê, divinité suprême en Basse-Egypte était représenté soit en tant que disque solaire, soit de temps en temps sous forme de faucon. Harmachis, dieu du soleil matinal et de la lumière de l'aube est un « Horus dans l'horizon » symbolisé par le sphinx[50]. Iao, Iao sont des éléments d'une invocation ou acclamation, des mots d'origine sémitique, désignant le caractère solaire et signifiant la radiation solaire[51]. La tête d'Apophis représente la victoire

46 Cf. BRELICH 1938, 131 ; NAGY 1942, 425-426.

47 VLASSA 1980, 137.

48 VLASSA 1980, 148-149 où la lecture *des mots est justifiée par leur présence fréquente dans toutes les deux formes ; cela est renforcé par BĂRBULESCU 1994, 165.*

49 Harmae = Harmachis, cf. VLASSA 1980, 144.

50 Pour un aperçu général voir DEMISCH 1977, 16-39, VLASSA 1980, 145 ; BĂRBULESCU 1994, 165.

51 VLASSA 1980, 149 ; BĂRBULESCU 1994, 165 ; cf. aussi BELLIDO 1967, 131 – l'adjectif qualificatif fréquent des dieux égyptiens. On le retrouve sur des amulettes magiques aussi, cf. HARRIS 1965, 93 ; concernant son rôle dans le culte des dieux égyptiens voir ERMAN 1934, 405 ; pour plus de détails loc. cit. 358-418.

sur les ténèbres[52] que nous pouvons voir sur le relief de Kunsthistorisches Museum sous forme d'un serpent au-dessous des pieds du sphinx. Le segment Re-Harmachis de l'épigraphe est un nom composé, signifiant une figure divine syncrétique[53]. Sur une inscription grecque, découverte près du Sphinx de Gizeh, nous retrouvons la même figure divine sous le nom de Hélios-Harmachis Theos Megistos[54].

Le sphinx était donc en Egypte le symbole des dieux solaires[55]. Ce symbole apparaît avec le nom de deux divinités représentées par le sphinx sur un emblème provenant d'une province avoisinante et datant de la même époque, issu d'un milieu semblable à celui des deux inscriptions de Savaria.

III. Dans la mesure où derrière la dédicace adressée aux sphinx de Savaria et d'Aquincum se cachent ces figures de dieux solaires égyptiens – une hypothèse que nous pouvons nous permettre –, il n'est pas surprenant que les constructeurs d'autel de jadis désignaient les deux divinités invoquées – de caractère plus complexe que le Sol romain, donc impossible à interpréter – par le nom de cette divinité, qui selon ses nombreuses représentations était un symbole explicite pour le monde contemporain, bien qu'existante en elle-même. Cela peut être mis en rapport avec le sphinx de Potaissa, dont la gaine comporte également deux noms de divinités, toutes les deux qualifiables de l'adjectif « majestueuse ». En ce qui concerne les figurations, je dois renvoyer au relief précédemment cité de Kunsthistorisches Museum, qui comporte lui aussi un disque solaire à côté des cornes de bélier. De même, sur la poitrine du sphinx de Potaissa nous pouvons déceler l'ancien symbole bien connu du soleil, le svastika[56], ce qui rendait encore plus évident à l'homme de l'époque ce que symbolisait

52 VLASSA 1980, 149-150 avec la littérature précédente.

53 NOLL 1955, Abb. 40 et p. 70.

54 OGIS II, 666.

55 Pour une argumentation détaillée voir VLASSA 1980, 142-147.

56 FORRER, 1914, 791-792

le lion ailé à visage de jeune fille[57]. D'après leur figuration et forme d'écriture, la datation des autels de Savaria peut être réduite au 3e siècle[58]. Bien qu'au 1er siècle après J.-C., l'existence du culte d'Isis à Savaria soit prouvée, ces deux dédicaces d'autel assez inhabituelles pourraient être reliées d'une manière plus convaincante éventuellement au grand flux d'immigrés orientaux de la fin du 2e et du début du 3e siècles[59]. Leur dédicace adressée aux figures du dieu solaire pouvait être l'expression des sentiments religieux des croyants habitant les lieux du culte des dieux égyptiens, qui tout en voulant s'aligner sur le mouvement religieux de l'époque, sur le culte de plus en plus intense du Soleil – Sol[60], étaient désireux de vénérer leurs divinités locales. Quant à la dédicace, elle contient sans ambiguïté l'adjectif qualificatif des figures divines impossible à interpréter, en respectant peut-être les règles du secret religieux relatif à la désignation des dieux.

Dans ce contexte, le rôle attribué – sous certaines conditions – aux sphinx représentés sur les autels peut paraître un peu exagéré. Comment le sphinx peut-il trouver sa place parmi les divinités hellénisées ou romanisées, en présence du culte empirique de Sol ? Et d'ailleurs, avait-il autant d'importance, que cela peut paraître aujourd'hui ?

Dans la seconde moitié du 4e siècle, c'est en mettant sur leur « drapeau » les images des dieux égyptiens que les sénateurs de la ville de Rome s'opposaient à la chrétienté, comme András Alföldi l'a démontré. Dans le cadre de ce mouvement, une série de

[57] A cet égard, l'illustration de l'amphore du Musée nationale de Parme, datant probablement de 380 à 60 avant J.-C., est très parlante : elle représente Œdipe avec le Sphinx, dont la tête est entourée d'une couronne de soleil rayonnante – voir KRAUSKOPF 1994, n. 66 ; voir aussi DEMISCH 1977, Abb. 46, 51, 57, 78.

[58] Cf. WESSETZKY 1961, 36.

[59] Concernant ce phénomène cf. par exemple FITZ 1972 ; de même que MÓCSY 1959, BARKÓCZI 1964, 257-356 et WESSETZKY 1961, 36.

[60] Cf. HALSBERGHE 1972.

médailles fut émise, dont le côté face représentait d'abord la figure du souverain, ensuite à cause d'une vive protestation chrétienne et impériale, les *sacra imagines* furent substituées par les images d'Isis, de Séraphis et de Sol-Séraphis. Cependant, le revers de ces médailles présentaient toujours les images d'importants dieux égyptiens admis par les Romains. Les onze types de revers se répartissent de la façon suivante : Isis se retrouve sur 6 revers de médaille – trois fois soit avec Harpocrates, soit avec le chien Sothis, ou encore avec Anubis ; Anubis est présent deux fois tout seul et une fois ensemble avec Isis ; un revers représente le Nil couché, un autre Harpocrates, tandis qu'un troisième le Sphinx, dans sa forme habituelle, c'est-à-dire avec des ailes et en mouvement[61]. Sur ces séries de médailles, émises pendant plusieurs décennies, et en rapport avec le vœu impérial de nouvel an, nous voyons donc le sphinx parmi les plus importantes figures de dieux « romains » - égyptiens, c'est-à-dire en tant qu'une figuration complexe de dieu solaire d'origine égyptienne[61], exempt d'*interpretatio.*

En connaissance de ces données nous arrivons peut-être plus près de l'interprétation des dédicaces solitaires et d'intention obscure de *Sp(h)incibus Augg(ustis), Sphinx(ibus)* de Savaria et de *Sphinc[ibus]* d'Aquincum. Dans mon exposé je n'ai pas eu l'intention de donner une réponse définitive aux questions, mais simplement faire intervenir dans l'étude des autels consacrés aux Sphinx d'Aquincum les possibilités offertes par celle du sphinx de Potaissa.

Ádám Szabó
Musée National de la Hongrie
Budapest

61 ALFÖLDI 1937, 13-19, XIII t. de 35 à 41, les deux premiers types de médaille constituent une exception. Sur le côté face de quelques types de médaille dont le revers représente le sphinx, se trouve une divinité couronnée de rayons – Sol-Séraphis.

Einige wichtige Fragen in Zusammenhang mit der Bearbeitung des Iseums von Savaria

Nach den vorläufigen Berichten und Teilpublikationen,[1] sind wir wegen der großen Menge der Materialien und der Bibliographie erst jetzt zu einem detaillierten Bearbeitung gekommen. Der enge Rahmen des Vortrags erlaubt uns nur zwei Fragekomplexe zu besprechen. Der erste ist das große Ausmaß und die Entstehung des heiligen Bezirkes, der andere ist der Fundort der Marmorrelief-fragmente und ihr vermuteter Platz an der ehemaligen Fassade.[2]

Das Ausmaß des Iseum savariensis ist 42 x 70 m. Es ist auch dann noch eine beträchtliche Größe, wenn sie mit dem Maßstab der erweiterten Stadt verglichen wird.[3] Jüngst hat Otto Sosztarics gezeigt, daß die Insulae der Stadt Savaria während der einzelnen Perioden der Römerzeit verschiedene Größe aufweisen konnten.[4] Auf die gleiche Schlußfolgerung kam die Untersuchung des Straßennetzes des Ruinengartens von Járdányi-Paulovics István.[5] Das Iseum ist aber verhältnismäßig leicht ins diese städtische Straßennetz zu stellen, denn es ist von Osten, Süden und Westen von basaltbedeckten Straßen-strecken umgegeben.

Der heilige Bezirk des Iseum savariensis ist ein organi-

1 WESSETZKY 1959; SZENTLÉLEKY 1960; SZENTLÉLEKY 1967; SZENTLÉLEKY 1978-79; SZENTLÉLEKY 1990.

2 Der Rekonstruktionsplan des Isis-Heiligtums von Savaria wurde durch Gyula Hajnóczy hergestellt.

3 Z.B. Der Vortrag von Otto Sosztarits whrend des jetzigen Symposiums.

4 BUÓCZ 1968.

5 SZENTLÉLEKY 1991.

scher Teil der architektonischen Reihe, die sich vom Apis-Kultplatz in Saqqara bis zur zweiten Hälfte des 2. Jahrhunderts, der Zeit der Bauarbeiten des Iseums in Savaria, herleitet.[6]

Am Bestattungsort des Apis-Stieres in Saqqara wurde das von den Ritualen bedingte Grundrißsystem entwickelt, das Einfluß auf die Heiligtümer des neuen Serapis-Kultes ausübte.[7] In Saqqara begrenzten Sphingen die breite Straße, die zum Tempel von Nektanebos I. führt. Löwen standen auf beiden Seiten des verzierten (arcusförmigen) Dromoseingangs, der vom Tempel 100 Meter entfernt lag. An seiner nördlichen Seite befinden sich die altägyptischen und hellenistischen Heiligtümer. Wichtig für uns ist das Gebäude des hellenistischen Heiligtums, weil sein System mit einer einzigen *Cellae* und zwei Eingangskapitellen versehen ist.

Vermutlich während der Regierung von Ptolemaios I., spätestens aber unter Ptolemaios III. wurde das Serapeum in Alexandrias Stadtviertel Rakotis errichtet.[8] Die Kulthandlungen spielten sich hier in einem großen, geschlossenen Hof ab.[9] Es ist noch nicht geklärt, wo im Hof das Isis-Heiligtum stand. Die Eingangsfassade ist reich vergliedert. Besucher wurden durch das Quervorhallen-System empfangen. Das große, hofähnlich geschlossene Gebiet wurde ein neues Charakteristikum der Serapis-Isis Heiligtümer der hellenistischen Zeit. Als Beispiel kann das Serapis-Isis Heiligtum auf der Insel Delos genommen werden,[10] die unter starkem ägyptischen Einfluß stand, weil die Ptolemäer sogar ihre Macht auf die Insel ausgedehnt haben. Auch hier wurde die *Cella* des Serapeums durch einen langen Dromos erreicht.

Der Weg zur Enstehung des Isis-Kultes in Savaria am Ende des 1. Jahrhunderts führt über Rom. Trotzdem konnte es Beziehungen auch nach Alexandria geben, weil die Seefahrt über

6 SZENTLÉLEKY 1960.
7 ROULLET 1972, 1-4.
8 ROULLET 1972, 30-31.
9 FRASER 1972, 268-270.
10 SALDITT-TRAPPMANN 1970, 117-172.

das ganz nahe Aquileia geführt hat,[11] worauf Inschriften hinweisen (Barbius Altar). Auf der Fassade des Iseums von Savaria sind die Gestalten von Serapis, Isis, Anubis und Harpokrates zu sehen.[12]

Zeitlich sind wir schon in der 2. Hälfte des 2. Jahrhunderts, als der heilige Bezirk in Savaria gebaut wurde. In der architektonischen Ausbildung des Iseums ist der Einfluß des großen Isis Heiligtums auf dem Campus Martius in Rom zu entdecken.[13] Vom Marmorrelief des Haterius-Grabes leitet ein Bogen zum Iseum, "Arkus" genannt. Der große Hof war durch eine Querhalle mit Säulen zugänglich.[14] Das Gebäude des Heiligtums findet sich in der Mittelachse des Hofes. In dem Iseum in Savaria ist der Eintritt auch an den zwei Seiten einer Querhalle zu finden. Von hier aus geht der Weg auf Treppen zu dem freistehenden Altar. Das Zentralheiligtum ist ein Megaron mit einer Vorhalle und mit einer zweigeteilten *Cella*.

Der Seitenraum und die Kapellen-Reihe sollen noch erwähnt werden. Die zwei zusammenhängenden Relieffrag-mente,[15] die im Archäologischen Museum von Rom und im Ägyptischen Museum von Berlin aufbewahrt werden, spiegeln seine äußere Ausbildung und Funktion wieder. Diese wurde in der neuesten Literatur gründlich ausgewertet. Die Szenen auf den zwei Wandgemälden in Herculaneum lassen Schlüsse auf die Riten vor dem Altar ziehen.[16] Über die Prozessionen in den Höfen sind wir aus den Metamorphosen von Apuleius in-formiert.[17]

Der besondere Wert des Isis-Heiligtums in Savaria ist nicht nur seine Ausdehnung, das System der verschiedenen

11 Vergl. BUDISCHOSVSKY 1977, 117-161.

12 SZENTLÉLEKY 1960, 14-18.

13 MALAISE 1972, 208-211.

14 ROULLET 1972, 25-26.

15 ROULLET 1972, Tafel XIII. – Das sogenannte Ariccia Relief, Rom –Museo Nazionale No. 77255; Ägyptisches Museum, Berlin – No. 16777.

16 TRAN TAM TINH, 1971, 29-41, no. 58-59, fig. 40-41.

17 Apuleius, Meth, XI. 1,4; XI. 7,1; XI. 23,1.

Hofniveaus und die große Säulenhalle aus Granit, sondern auch das Marmorfries mit den Szenen. Entsprechend den Daten der Freilegung hat Gyula Hajnóczy den theoretischen Rekonstruktionsplan angefertigt (1960).[18] Der Plan wurde 1963 erstellt. (Prof. Gyula Hajnóczy lebt nicht mehr.) Nach dem Erscheinen vieler EPRO und anderer Publikationen wurden mehrere neue Angaben und Erkentnisse bekannt. Deswegen halte ich es für sehr wichtig, einige Angaben und Zusammenhänge, die schon während der Ausgrabung klargeworden sind, hier zu betonen.

Zuerst sollen einige Dinge über die Ausgrabungen gesagt werden. Der Auslöser für die Ausgrabungen waren Bauarbeiten: das Depotgebude der Vasmûszaki Firmen in der Straße Tököly sollte ursprünglich erneuert werden. Als aber die Erdarbeiten begannen, wurden die Ausgrabungen des Fundaments verwirklicht, nämlich einige Steinblöcke wurden gefunden, deren Material stark granulöser, harter Kalkstein ist. Er wird lokal Halbmarmor oder stark granulöser Marmor genannt. Die wurden als römische Bausteine identifiziert. Während der Freilegung wurden die Dimensionen eines Zentralheiligtums geklärt. Es wurde auch festgestellt, daß es ein Iseum mit zweiteiliges Megaron und mit einer vorderen Querhalle ist. Bei den Grundwänden der inneren *Cella* stand möglicherweise ein Podium. ähnliche Gebäude und Funde sind vom Isis-Heiligtum von Ras el-Soda in Alexandria bekannt.[19] Auch Marmor Kultstatuen wurden in Alexandria gefunden. In Savaria ist aber der Grundbau der vermuteten podiumteilenden Wände schwach, daher konnten hier nur leichtere Statuen für die Prozessionen aufbewahrt werden, die gelegentlich bekleidet wurden.

An der nordöstlichen Fassadenecke des Heiligtums waren die Schichten während der Ausgrabungen noch unberührt geblieben, und das Fundament des 4 x 4 m großen Altars war noch gut erkennbar.[20] Der Bereich rund um den Altar ist gle-

18 HAJNÓCZY-SZENTLÉLEKY 1964, 129-136.

19 SZENTLÉLEKY 1960, 11-12. Fig. 6.

20 SZENTLÉLEKY 1965, 97-114; SZENTLÉLEKY 1967, 457-466.

ichzeitig der tiefste Punkt des heiligen Bezirkes. Die Marmorreliefs mit Darstellungen sind mit einer Ausnahme auf diese Schicht gefallen. Man fragt sich, woher die Marmorblöcke mit den Szenen auf diesem untersten Niveau herkommen konnten.

Auf der östlichen Seitenwand des Grundes der Vorhalle lief eine starke Bruchlinie vom Fundament bis zum Laufniveau der Vorhalle. Auf beiden nördlichen und südlichen Seiten der Fassade laufen Antenmauern nach Osten. Diese Stützmauern schützen den kleineren Raum um den Altar.

Bis 1960 kamen, während der planmäßigen Ausgrabungen die Reste des heiligen Gebäudes zum Vorschein. Der schon früher erwähnte Hof wurde besser gestaltet, dessen betontestes Element das Zentralheiligtum war. Vor der nördliche Ecke der östlichen Fassade des Heiligtums erstreckte sich der tiefere Raum, wohin die Reliefs mit einer Ausnahme stürzten. Dieser eine Marmorblock hat wegen des Inhaltes der Szene eine besondere Bedeutung (Tf. 32/1).[21] Auf seiner Oberfläche stehen drei Gestalten. Sie tragen Götterattribute. In der Mitte ist ein Priester als Anubis verkleidet und mit einer Hundekopfmaske zu sehen, der eine Tunica trägt (Tf. 32/2). Auf seiner rechten Seite ist Isis dargestellt. Eine Schlange kriecht auf ihren Armen. Auf der linken Seite des Anubis-Priesters nimmt Serapis Platz. Er trägt einen Kalathos auf seinem Kopf und hält seinen Stab in der rechten Hand. Weil der Priester, der eine Tunica trägt, eine Hundekopfmaske[22] in der Rolle des Anubis hat, zeigt die südliche Strecke der Fassade die Lieder des von Apuleius gut bekannten Isis Prozessionen.

Auch die Darstellung des Frieses des nördlichen Teils der Fassade paßt gut zu diesem thematischen Programm. Diese Reliefblöcke sind nur aus der Zeit der Ausgrabungen – 1955 – bekannt. Schon mehrere Forscher haben die inhaltlichen Fragen studiert. Die von der antiken Literatur so berühmte Beschreibung von Dio Cassius ist für uns sehr wichtig, nach der auf der Fassade des Iseums auf dem Campus Martius in Rom

21 MÓCSY-SZENTLÉLEKY 1971, 96, Kat.Nr. 74.

22 siehe Abb. 1.

die Isis-Gestalt, auf einem Hund galoppierend, zu sehen war. Die Geschichte des Heiligtums in Rom ist sehr wechselreich. Vom Gesichtspunkt des Iseum savariensis her sind nicht seine Anfänge, sondern die kaiserzeitlichen römischen Bauarbeiten wichtig. Unter der Regierung von Tiberius wurde das Heiligtum zerstört, unter Domitian es reichlich ausgebaut. Zwischen den beiden Kaisern stand schon das Isis geweihte Heiligtum, weil auf den römischen Münzen des Kaisers Caligula die Fassade des Heiligtums auf dem Campus Martius zu sehen ist, und zwar mit der Gestalt der Isis, die auf einem Hund galoppiert.[23]

András Alföldi hat uns in der Mitte des 20. Jahrhunderts auf das Weiterleben des Isis-Kultes im 4. Jahrhundert aufmerksam gemacht,[24] für Savaria und sein Iseum tat dies im Zu-sammenhang mit dem Silbermünzenfund in dessen Nähe István Tóth.

Ein weiteres Bild kann nur mit der Hilfe der Schilderung der ganzen Umgebung des Bildes beschrieben werden. Es gibt drei solche Punkte auf dem Zentralheiligtum und dessen Umgebung, wo der Fries mit der Szene plaziert sein konnte:

1) auf der Seite des Altars
2) auf der Balustrade der Vorhalle
3) auf der Fassade des Zentralheiligtums

Auf Grund der Ausgrabungen wurde festgestellt, daß ein 4 x 4 m großes Altarfundament vor der Ostseite des Heiligtums stand. Auf den zwei Gemälden von Herculaneum befindet sich der Altar vor der Fassade. Auf dem einen ist der Altar ein niedriger Bau, und das Ritual einer Wasseropferung dargestellt. Auf dem anderen Bild erscheint die wilde Tanzszene eines Brandopfers. Der Altar ist hier höher, und der an seinen vier Seiten betonte Akroterschmuck, der auch gegen das Zerfallen schützt, erleichtert das Aufschüren der Glut. Das Altarfundament in Savaria entspricht diesen Aufgaben.

Zurück zum Fries mit den Darstellungen. Es ist wegen zwei

23 ROULLET 1972, 1-5.

24 ALFÖLDI 1937.

Ortsangaben unmöglich, daß sie zum Altar gehörten. Die Folge der Reliefs der aneinandergrenzenden polierten Marmor-blöcke macht ca. 4,5 m aus, und erst dann folgt das beschrie-bene Feld. Also wäre das 4,5 x 4 m große Altarfeld viel kleiner als das Feld, das gebraucht wurde. Gleichzeitig wären die 40 cm hohen Mamorblöcke auch sehr schwer. Bei der Ausgrabung wurde das Altarfundament eben deswegen gründlich untersucht, und festgestellt, daß es keine Gründung in beträchtlicher Tiefe hat.

Die zweite Möglichkeit wäre die Balustrade der Vorhalle. Es ist deswegen bemerkenswert, weil auf dem Gemälde mit der Opferszene in Herculaneum, die Querhalle des Iseums an der Fassade vor dem Altar, gegen das tiefere Niveau, an beiden Seiten mit Steinbalustraden begrenzt ist. Was jedoch noch störender ist, ist daß ihre Oberfläche figürliche Darstellungen zu haben scheint, und horribile dictu sieht es so aus, als ob ein Tierkopf vorhanden wäre. Auf beiden Seiten der Balustradenbrüstungen liegt je eine Sphinx. Die zwei kleinen Wandgemälde sind im Archäologischen Nationalmuseum Neapel ausgestellt.

Das wahrscheinlichste System der Anordnung des Reliefs in Savaria entspricht der jetzigen Aufstellung an Ort und Stelle. Die architektonischen Gesichtspunkte bestimmen primär das rekonstruierte Bild des ehemaligen Zustandes, während die ehemalige Situation uns aber durch die gefundenen Steindenkmäler übermittelt ist. Gyula Hajnóczi hat, als Archäologie-Professor der Römischen Architektur, die Struktur des ehemaligen Gebäudes rekonstruiert, und er hat als Architekt, die heutige Form für den Ausbau des Ruinengartens gefunden. Zusammen mit dem Archäologen, der die Ausgrabungen leitete, hat er die Pläne die damaligen vorwärtsweisenden Neuerungen des Nachbaus benutzend – so bearbeitet, daß wir heute, 40 Jahre später ruhig und mit genügendem Grund an der Bewahrung dieses ausergewöhnlichen Monumentes für die nächsten Jahrhunderte weiterarbeiten können.

Tihamér Szentléleky
Budapest

APULEIUS "PANNONICUS"*

The life-span of Apuleius of Madaura (cca. 125 +/-50?-70? years) correlates exactly with the diffusion and florescence of the Egyptian cults in Pannonia. It can be proved, that the most important places of his life – North-Africa, The Balkan, Asia Minor, Syria, Rome – impressed directly the Pannonian religious practices.

The paper compares the moments of the actual religious practices written down in the texts in the Metamorphoses XI. with the monuments found in Pannonia and the conclusions you can trace from them. It concludes on this base, that the Apuleius text, and the Pannonian reliques correlaters exactly. Thus the characterisca of the ground-plan of the Savaria Iseum, the themes and iconography of its sculptural design, the details of the ceremony reconstructed and their accessory, details of the great cultic procession, the phraseology of the epigraphic monuments all corresponds to the description of the Isis religion and the initiation rites in the Metamorphoses.

The results are based beside the works of Aladár Dobrovits, Vilmos Wessetzky, László Kákosy, J. Gwyn Griffith, Günter Grimm and Michel Malaise on the researches partly pubished of the present author, who has collected the whole Panno-nian reliquia of the Egyptian cults.

István Tóth
Univ. Janus Pannonicus
Pécs

The lecture appeared in Hungarian in: Savaria 25/3, 2000-2001, 192-197.

Monuments égyptiens en Hongrie

En 47 av. J.-C., lorsque Jules César a occupé Alexandrie, la bibliothèque d'Alexandrie a été mise en cendre et d'après la tradition, sept cent mille volumes ont péri. Les oeuvres portaient sur l'histoire égyptienne, sur les actes des pharaons, c'est ici qu'a péri entre autre l'Histoire égyptienne en trente volumes du prêtre égyptien Manéthon, ouvrage écrit en grec sur la demande de Ptolémée II Philadelphe. Manéthon a écrit son histoire sur la base des oeuvre gardées dans les bibliothèques et archives des temples et s'est occupé non seulement des événements d'antan, mais également de la religion et des coutumes des Égyptiens. Il est vrai que toutes les grandes bibliothèques, ainsi celle d'Alexandrie, possédaient des copies et dans ce cas, celles-ci étaient également à Alexandrie, à Sérapiéon, mais dans les siècles suivants ce temple a été fermé, détruit et livré aux flammes en 391. Ainsi on péri les oeuvres échappées à la catastrophe de 47 av. J.-C. Par son décret de 391, l'empereur Théodose a fait fermer tous les temples païens de l'empire, dont les temples égyptiens, et la pratique de l'écriture hiéroglyphique s'est définitivement terminée. Vers 450, lorsque personne n'écrivait plus ni ne lisait les anciens textes égyptiens, tout ce que les Égyptiens eux-même avaient écrit en grec sur leur pays pour le faire connaître à leurs occupants, a également disparu.

Cependant, les oeuvre des auteurs classiques grecs et latins s'intéressant à l'Égypte ont subsisté à Rome, et à Byzance. Les livres de l'Ancient Testament, comme p.ex. la Genèse ou l'Exode et d'autres ont gardé des fragments de l'histoire politique de l'Égypte, et dans leurs écrits les prêtres de l'Église citaient souvent les oeuvres de Manéthon pour justifier l'authenticité de l'Ancient Testament. C'est par ce moyen que la division de l'histoire de l'Égypte en 30 dynasties nous est restée, division que l'égyptologie moderne a acceptée.

La tradition classique hellénistique et romaine a souvent évoqué la religion égyptienne qui était pour eux singulière et attirante. Le culte de la déesse Isis, d'Osiris et d'Anubis s'est répandu dans l'empire romain et a développé les rites présumés de l'époque des pharaons. Ainsi s'est répandue la légende d'Osiris (De Iside et Osiride), légende dont les sources égyptiennes initiales parlaient avec beaucoup moins de détails.

Les épisodes connus de la Bible: la traversée de la Mer Rouge, la vente de Joseph, puis son activité dans la cour du pharaon, l'exposition sur le Nil du nourisson MoÏse dans un panier et son adoption par la fille du pharaon – tous cela a survécu comme souvenirs vivants à l'époque du Moyen-Age et dc la Renaissance.

Les marins de la Palestine et de la Syrie ont été les premiers à visiter l'Égypte, mais il ne nous pas parvenu de textes d'eux sur ce qu'ils ont vu et observé. De leur venue témoignent les peintures murales des tombeaux égyptiens (p.ex. celles des tombeaux de Khnoumhotep à Beni Hassan, autour de 1900 av. J.-C., ou de Sebekhotep à Thebes, autour de 1400 av. J.-C.). C'étaient d'abord les impressions de voyage des Grecs qui ont fourni des informations sur le monde de la vallée de Nil dans l'Odyssée, Homère a décrit l'irruption dans le Delta des pirates grecs, massacrés ou faits prisonniers par les Égyptiens. Toutefois, des mercenaires et de paisibles commercants grecs ont aussi séjourné ici: les pharaons de la XXVIe dynastie ont engagé des mercenaires grecs de l'Asie Mineure d'abord dans les luttes contres les Assyriens, puis dans celles contre les Perses. C'est dans ces temps que se sont établis les commercants grecs à Naucratis dans le Delta ou à Éléphantine près des garnisons militaires, bénéficiant de l'appui du pharaon. Lorsque Cambyse a occupé l'Égypte en 525 av. J.-C. et qu'il est devenu le maître de toute l'Asie Mineure, les Grecs des rives ioniennes et des îles proches ont été ses sujets.

Le voyageur-historien grec Hérodote de Halicarnasse est arrivé en Égypte en 450 av. J.-C., après avoir lu tout ce que les

Grecs avaient noté sur ce pays et il n'a pas caché sa sympathie à l'égard de ses habitants. Dans son deuxième livre, il a fait des descriptions pleines de fantasie; ces descriptions tirent leur origine partiellement d'informations inexactes, mais en raison de leurs détails précieux, les égyptologues utilisent son livre même aujourd'hui. Ce sont en premier lieu ses notes sur la vie quotidienne, sur la religion et sur l'environnement naturel qui sont les plus riches parmi ses écrits. C'est par lui qu'on peut faire la connaissance des traits les plus importants de la vie des Égyptiens dont ni les figurations, ni les textes ne donnent d'informations exactes. Il a décrit leurs grandes fêtes religieuses et le culte rendu aux bêtes honorées en tant que saintes.

Les auteurs grecs et romains après Hérodote ont écrit leurs comptes rendus sur l'Égypte hellénisée. Contemporain de Jules César, Diodore de Sicile a égelement visité l'Égypte. On ne peut pas distinguer dans son oeuvre ce qu'il a puisé chez Hérodote et ce qu'il a entendu des Égyptiens. Strabon est arrivé en Égypte 50 ans après Diodore, vers 30 av. J.-C., lorsque l'Égypte est devenue une province romaine. Avec l'aide de son ami gouverneur, Aelius Gallus, il a parcouru le pays, et a fait, dans son ouvrage intitulé *Geographica*, des descriptions distrayantes et exactes (des fêtes populaires à partir d'Alexandrie jusqu'à Canope, du culte à Crocodeilopolis, du Serapeum). Prêtre d'Apollon à Delphes, Plutarque, biographe grec, a voyagé en Égypte vers 100 apr. J.-C., lorsque la copie de l'oeuvre de Manéthon existait encore à Alexandrie, et il a écrit l'histoire de la légende d'Osiris puisant à celle-ci aux traditions locales.

Les empereurs romains eux-mêmes faisaient avec plaisir des voyages en Égypte: Hadrien et Septime Sevère ont même gravé leur nom sur les colosses de Memnon. De nombreux empereurs ont construit les sanctuaires des anciens dieux. Nous apprenons de Tacite que Germanicus a visité en 19 apr. J.-C. les temples de Thèbes, où un prêtre égyptien lui a traduit les textes hiéroghlyphiques et c'est également de Tacite que proviennent nos information sur les impôts dont les Romains ont frappé les nations occupées. Pour orner leur capitale, les empereurs emportaient à Rome et à Byzance des monuments égyptiens, dont les inscriptions n'ont cessé d'intriguer la fantaisie des savants. Dès le début du XVIIe siècle, Athanasius Kircher a tenté de résoudre l'écriture hiéroglyphique. C'est néanmoins aux voyageurs qu'on doit d'avoir fait durer l'intérêt à l'égard du passé de l'Égypte considérée comme mystérieuse.

Pendant les siècles du Moyen-Age, seules les descriptions des croisades (1099-1270) ont donné des rensegnements sur l'Égypte, mais elles pouvaient à peine pénétrer dans ce pays musulman et n'allaient guère au-delà des environs du Caire. La plupart des pélerins ne rendaient compte que la région du Delta et des pyramides, considérant ces dernières comme les entrepots de blé du Joseph biblique. Ils cherchaient en premier lieu les monuments chrétiens, de même qu'il tenaient l'Égypte seulement pour une étape du voyage vers la Terre Sainte. Finalement, la possibilité de faire des grandes voyages s'est ouverte au XVIIe siècle. Des religieux capucins, dominicains et jésuites se sont rendus au Caire avec l'intention d'évangéliser. En 1672, le dominicain Vansleb a visité le pays, il est parvenu jusqu'à la Haute-Égypte et a acheté de vieux manuscrits. Tout comme les autres religieux, il s'est émerveillé des anciens monastères coptes, mais il a aussi consacré son attention aux monuments plus ancestraux: il a été le premier Européen à décrire les ruines d'Antinoé, ville romaine qu'Hadrien avait fait bâtir pour son favori, Antinoüs. L'époque des grandes découvertes est enfin venue quand Napoléon, accompagné de ses troupes, puis les représentants des différentes sciences ont examiné l'Égypte et ont préparé des descriptions détaillées du pays, dans 9 volumes de texte et 11 volumes de tableux. Ces descriptions ont servi de base au développement de l'égyptologie. D'un jour à l'autre, l'Égypte est devenue à la mode en Europe. Parallèlement, le pillage et le vol des monuments égyptiens ont commencé. Le dommage a été énorme, mais les monuments enlevés ont servi de matériel pour déchiffrer l'écriture hiéroglyphique.

Les pillage avaient commencé des l'époque égyptienne classique (tombes royales), et avaient continué pendant les empires romain et byzantin: les empereurs avaient emmené des obélisques, des sphinx, des statues dans leurs capitales ou seulement dans leurs villas, comme p.ex. Hadrien ou Dioclétien. Précédemment, les Perses avaient également emporté des statues égyptiennes à Persepolis. Les monuments avaient été ravagés par les coptes (transformation des sanctuaires païens en églises chrétiennes) et par les ermites (occupation de vieux tombeaux, mutilation des statues, grattage des détails des reliefs et des peintures murales, ce qui constitue également une espèce de pillage). Même une liste avait été préparée en arabe des lieux où l'on pouvait trouver des trésors en pratiquant certains rites magiques. Cette liste a survécu jusqu'au début du XXe siècle et a occasionné beaucoup de dégats (avec des marteaux contre les stèles).

Au XIXe siècle, de vrais aventuriers ont parcouru l'Égypte et ont pillé les tombeaux et les monuments de manière systématique. Les consuls des pays européens ont fait fouilles eux-mêmes ou les ont fait par leurs agents, ont acheté des objets égyptiens ou simplement ont donné charge de les dénicher et de les enlever. La naissance des grandes collections européennes est un chapitre long et passionnant de l'histoire de l'Égypte redécouverte. En même temps, c'était ce même siècle ou l'étude des documents originels: fragments architecturaux, statues, papyrus et reliefs a permis de déchiffrer l'écriture et de faire ainsi parler l'Égypte antique.[1]

En 1799, lors des travaux de construction à Rosette dans le Delta, l'attention de l'officier français Pierre Bouchard a été attiré par une pierre noire avec inscription, transportée à Alexandrie. C'était une stèle constituée de trois parties, sur laquelle il y avait en haut une texte hiéroglyphique gravé, au milieu des signes cursifs évoquant vaguement l'écriture arabe, et en bas une inscription grecque. Les linguistes ont traduit le texte grec et ont constaté qu'il s'agit d'une copie du décret de Ptolémée V de 196 av. J.-C. et ils ont supposé que l'inscription grecque était la traduction des deux textes supérieurs. Tombée en mains anglaises comme prise de guerre et gardée aujourd'hui

[1] Berlin 1990.

dans la British Museum, la stèle avait heuresement été copiée. Elle est parvenue ainsi à Paris, où un capitaine nommé Champoléon l'a transmis à son neveu, J.F. Champollion, qui, en 1822, a déchiffré l'écriture hiéroglyphique. Cela a donné naissance à la science de l'égyptologie. Les bases de l'égyptologie se sont développées sous peu dans de nombreux pays de l'Europe: des experts ont étudié les monuments et se sont rendus en Égypte pour diriger des fouilles. Chaque savant européen était en même temps patriote et voulait enrichir son propre pays (Lepsius, Rosellini, Maspero, etc.).

Par suite de la capmpagne de Napoléon, ainsi que la parution des volumes imposants de l'oeuvre intitulée *Description de l'Égypte* et du transport des monuments, ont a pu observer une vague de style égyptisant en Europe. Les arts décoratifs (bijoux, porcelaines, meubles), l'architecture, la peinture, la sculpture, la musique et la littérature – surtout dans la deuxième moitié du XIXe siècle – ont utilisé des motifs égyptiens. Les éléments égyptiens se sont mélés au classicisme, au néogothique et à l'empire et, surtout au début de XXe siècle, à l'Art Nouveau et aux différents tendances artistiques cherchant du nouveau.[2]

Au XIXe siècle, la Hongrie n'était pas en mesure de s'associer aux recherches égyptiennes, toutefois les résultats mis au jour par les recherches archéologiques et historiques et par les journaux de voyage n'ont pas échappé à l'attention des savants. Des voyageurs hongrois sont parvenus en Égypte dès les XVe et XVIe siècles, surtout des prêtres et des moines qui, se rendant à la Terre Sainte, ont débarqué à Alexandrie, puis ont continué leur voyage par Le Caire.[3] En 1483, l'archidiacre János Lázoi a même visité les pyramides et a fait un voyage en bâteau sur le Nil. György Huszti Raszinyai a été emmené en esclavage par les Turcs en 1532 et bien que cultivé, il a choisi d'être corniste et non secrétaire. Après ses fuites répetées, il est entré en service du sultan, mais n'a pas pu rentrer dans sa patrie. Finalement, il s'est

2 Wien 1994.

3 HORVÁTH 1985.

embauché sur un navire comme matelot, et ainsi il est parvenu en Égypte. D'après son compte rendu, il a passé deux heures dans la pyramide de Chéops, puis a grimpé en haut de la pyramide et a amusé ses camarades par le sonnerie du cor. Il s'est évadé à travers les Indes pour rentrer dans sa patrie et s'est installé en 1542 à Pozsony. Parmi les voyageurs, il y avait des diplomates, comme baron Ferenc Tóth, envoyé de Bethlen et des historiens, comme István Benczédi Székely, dont l'oeuvre publiée en 1559 traite longuement l'histoire égyptienne.

Au tournement des XVIIIe et XIXe siècles, plusieurs historiens hongrois éminents ont consacré des ouvrages remarquables à l'histoire de l'Égypte. Pál Bertalanffy (1757) a écrit une oeuvre sur la religion intitulée Aegyptus, János Molnár (1760) a fourni des informations sur les édifices (pyramides, tours solaires, labyrinthes), et a mentionné un homme nommé Kirker qui connaissait l'écriture hiéroglyphique (Athanasius Kircher), József Gvadányi a écrit sur Osiris et un bâtiment "idole": le Sphinx. Oeuvre éminente publiée en 1803, le livre de Sámuel Decsy, intitulé "Egyiptom ország historiája" (L'histoire du pays d'Égypte) décrit les résultats de l'expédition française de même que ses connaissances sur la momification.

Dans la première moitié du XIXe siècle, se sont surtout les journeaux qui ont rendu compte des sensations scientifiques: de Champollion, d'Alexandrie, des monuments architecturaux. En 1832, on pouvait lire dans l'ouvrage intitulé "Közhasznú ismeretek tára" (Collection des connaissances utiles) à l'entrée Égypte sur les anciennes coutumes égyptiennes, sur les momies, sur les bijoux, sur les bêtes. Dans la deuxième moitié du siècle, des oeuvres historiques importants ont été publiées par Ferenc Ribáry (Az ókor története = Histoire de l'Antiquité), par Gyula Pasteiner (Mûvészet = Art), par György Polgár (Régiségek = Antiquités) et Vilmos Petz a rédigé l'Encyclopédie de l'Antiquité, la première en Hongrie, publiée en 1902. Les journaux les plus connus, comme Budapesti Szemle (Revue de Budapest) et Pester Lloyd publiait systématiquement des articles de vulgarisation. Tous ces comptes rendus, impressions de voyage, articles historiques ont incité non seulement les prêtres et les moines à faire des voyages, mais également des laïcs aisés et des aristocrates. En 1840, Lukács Gyertyánffy a gravé son nom sur plusieurs monuments égyptiens: sur la statue d'Amon dans le sanctuaire du temple rupestre d'Abou Simbel et sur un des pylones de Ramesseum.[4] En ce temps, il vivait en Égypte de nombreux Hongrois: des consuls, des moins, des commerçants et également un médecin de renom. Ils ont contribué à ce que les voyageurs hongrois ont acheté des objets d'art égyptiens et les rapportant en Hongrie, les ont offerts au Musée National, comme p.ex. le comte Antal Eszterházy, qui a acheté et a offert en cadeau de nombreux trésors artistiques: les volumes des années 1871 et 1901 de la revue Archeológiai Értesítõ en donnent un compte rendu fidèle. D'après la tradition orale, Eszterházy a acheté en Égypte des tubercules de lotus du Nile pour 13 kilos d'argent et les a fait implanter dans le lac de Hévíz.

Aujourd'hui, on ne peut plus déterminer exactement la date à laquelle les premiers objets égyptiens sont arrivés en Hongrie. Toutefois, il est certain qu'en 1825, on avait enregistré plus de 100 examplaires dans le catalogue du Musée National et ce nombre s'est accru sans cesse durant le siècle.[5] Dans les années 60, György Sipos, jardinier hongrois du pacha Fouad, a demandé à son maître, en tant que récompense de ses services, une magnifique statue calcaire et il l'a remise au Musée National. Ferenc Pulszky a acheté de la collection d'Alexandre Renoir la statue d'Imhotep de l'Abbaye Saint-Denis à Paris, statue en bronze avec incrustation d'argent. En 1885, Károly Markstein, ancient éléve du lycée de Pápa, a acheté en Égypte un très beau sarcophage de momie avec une momie orné, originaire d'une époque postérieure, pour fonder la collection égyptienne de sa ville.[6] La pièce est encore aujourd'hui au Lycée réformé de la ville. Miksa Déri a acheté en 1895, sur la base de l'expertise de

4 Péter Gaboda, égyptologue, traite l'histoire des voyageurs hongrois qui ont visité l'Égypte

5 DOBROVITS 1939. Tout récemment VARGA 1994.

6 VARGA-LÁSZLÓ-KOVÁCS, 1987.

l'illustre égyptologue allemand, E. Brugsch, les objets collectés pour le consul austro-hongrois Cischini, objets qui ont été placé au Musée Déri à Debrecen et dont son fils, Frigyes Déri a dressé un catalogue.[7] Les deux sarcophages de momie de la collection proviennent du voyage en Égypte de l'archiduc Francois-Ferdinand. Au tournant du siècle, Bonifác Platz, abbé de Zirc, savant et collectionneur d'objets d'art a visité Égypte plusieurs fois avec des professeurs et ses disciples, et y a acheté, en connaisseur, une collection intéressante.[8] Lors de son premier voyage, il a été accompagné par Zsolt Beöthy, homme de lettre, professeur d'université qui, plus tard, a effectué lui-même plusieurs voyages en Égypte et a collecté un ensemble précieux d'objets d'art.[9] Après son décès, cette collection a malheuresement été vendue par ses héritiers à l'Université d'Uppsala à cause de la parcimonie de l'université hongrois. Les monuments égyptiens parvenus en Hongrie ont été enrichis par les fouilles menées entre 1907 et 1909 à Charouna et Gamhoud en Égypte Moyenne,[10] financées par le commerçant hongrois Fülöp Back et dont le matériel (reliefs de temple, sarcophages en bois contenant des momies et des ornements de l'époque ptolémaïque et de l'époque romaine) a été réparti entre les musées de Vienne et de Budapest et en petite partie celui de Cracovie.[11]

Ede Mahler, fondateur de l'égyptologie hongroise, a lutté pour unir l'ensemble des monuments égyptiens en Hongrie en une seule collection. Toutefois, la Collection égyptienne au Musée des Beaux-Arts n'a été réalisée qu'en 1934, après trois décennies d'efforts.

Outre son activité de niveau international déployé dans les domaines de la chronologie, des sciences mathématiques et de

7 DÉRI 1922, 300-313.

8 HORVÁTH 1985, 60-65: B. Platz, Önéletrajzi adatok (Données autobiographiques)

9 MAHLER 1913.

10 KAMAL 1908.

11 Récemment: GYŐRY 1997.

l'astronomie, Ede Mahler a participé en tant que collaborateur de la Collection d'Antiquités du Musée National à l'exploration des monuments de Pannonie.[12]

Il est bien connu qu'il a longtemps travaillé à Intercisa et c'est lui qui a trouvé dans la rue Perc à Óbuda un monument avec inscription égyptienne et l'a considéré comme monument religieux égyptien de l'époque romaine pannonienne. Ses articles dans la revue Archeológiai Értesítõ rendent compte de ses recherches. Ses disciples: Aladár Dobrovits et Vilmos Wessetzky se sont occupés des monuments égyptiens de la Pannonie en tant que partie de leurs travaux d'égyptologique. Par l'analyse historique et d'histoire des religions, par ses études sur les cultes égyptiens à Aquincum, sur des monuments égyptiens et égyptisant de la Russie du sud, de la région de la Mer Noire entre le premier siècle av. J.-C. et le troisième siècle, Dobrovits a contribué à mettre en lumière la propagation et l'interprétation des cultes des dieux émanant de l'Égypte.[13] Vilmos Wessetzky a beaucoup de mérites dans les recherches sur le culte d'Isis à Savaria et sur les trouvailles égyptiennes. Il a publié le matériel des monuments dans un volume particulier, oeuvre de base des chercheurs des décennies passées, du temps présent et certainement de l'avenir aussi, bien que depuis cette époque, de nouvelles recherches aient été réalisées et que des résultats nouveaux soient nés concernant le culte et l'Iseum de Savaria.[14] C'est le point ou se rencontrent les intérêts des archéologues et des philologues étudiants la Pannonie et celui des égyptologues. En effet, nous trouvons en Pannonie des objets égyptiens originels de même que des statues romaines de dieux de style égyptisant ou des statues de dieux hellénistiques en pierre ou en bronze, et même des sculptures en pierre avec des scènes égyptisantes ou d'origine égyptienne. Les monuments des inhuma-

12 Sur l'activité et la liste des oeuvres de E. Mahler v. HORVÁTH 1985, 92-142.

13 P. ex.: DOBROVITS 1943, DOBROVITS 1954, etc.

14 WESSETZKY 1961, etc.

tions de momies laissent supposer que la population de cette province a subi l'influence non seulement des dieux égyptiens et de leurs cultes, mais également des coutumes égyptiennes d'inhumation. Il serait souhaitable de procéder à la révision et à l'authentification de la découverte des trouvailles éparses égyptiennes originelles, considérées comme appartenent aux cultes égyptiens. La coopération des chercheurs de diverses spécialités permettra d'explorer les détails du caractère romainisé de la culture égyptienne en Pannonie.

Edith Varga
Musée des Beaux-Arts
Budapest

ABRÉVIATIONS – BIBLIOGRAPHIE

Manuels, encyclopédies, périodiques

AA = Archäologischer Anzeiger, Berlin.

Acta Arch = Acta Archaeologica Academiae Scientiarum Hungaricae, Budapest.

Acta Ant. = Acta Antiqua Hungarica, Budapest.

AEM = Archäologisch-epigraphische Mitteilungen aus Österreich, Wien.

AF = Ägyptologische Forschungen, Glückstadt - Hamburg - New York.

AMSIA = Atti e memorie della Societŕ istriana di archeologia e storia patria, Parenzo-Trieste.

ANRW = Aufstieg und Niedergand der Römischen Welt, Berlin-New York.

Arch Ért = Archeológiai Értesítő, Budapest ab 1869.

ARR = Archeološki radovi i rasprave (Acta et dissertationes archaeologicae), Zagreb.

AT = Antik Tanulmányok, Budapest.

AV = Arkedoski Vestnik, Ljubljana.

AW = Antike Welt, Basel.

BASD = Bullettino di archeologia e storia dalmata, Split.

BdE = Bibliothèque d' étude, Le Caire.

BIFAO = Bulletin de l' Institut Français d'Archéologie Orientale, Le Caire.

BMC,IV, 1940 = H. Mattingly, Coins of the Roman Empire in the British Museum, IV, Antonius Pius to Commodus, London 1940.

BMHBA = Bulletin du Musée Hongrois des Beaux-Arts, Budapest.

BONNET = H. Bonnet, Reallexikon der ägyptischen Religionsgeschichte, Berlin 1952.

BSFE = Bulletin de la Société Française d' Égyptologie, Paris.

Bud. Rég. = Budapest Régiségei, Budapest.

CIL = Corpus Inscriptionum Latinarum, Berlin ab 1863.

CIL 3 = Th. Mommsen, Corpus inscriptionum latinarum, consilio et auctoritate Academiae litterarum regiae Borussicae editae, vol. III pars posterior: Inscriptiones Asiae provinciarum Europae Graecarum Illyrici latinae (Suppl. 1902-1903), Berlin 1873.

CIL 5 = Th. Mommsen, Corpus inscriptionum latinarum, consilio et auctoritate Academiae litterarum regiae Borussicae editae, vol. V: Inscriptiones Galliae Cisalpinae latinae, Inscriptiones regionis Italiae decimae comprehendes, Suppl. Pais 1884-1888, Berlin 1872.

CIL 9 = Th. Mommsen, Corpus inscriptionum latinarum, consilio et auctoritate Academiae litterarum regiae Borussicae editae, vol. IX: Inscriptiones Calabriae, Apuliae, Samnii, Sabinorum, Piceni Latinae, Berlin 1883.

CommArchHung = Communicationes Archaeologicae Hungariae, Budapest.

CSIR = Corpus Signorum Imperii Romani.

DAREMBERG-SAGLIO = Dictionnaire des Antiquités grecques et romaines, ed. Ch. Daremberg - M. E. Saglio, Paris 1908.

EPRO = Etudes préliminaires aux religions orientales dans l' Empire Romain, Leiden.

ÉT = Études et Travaux, Warszawa.

FORRER 1914 = R. Forrer, Reallexicon der prähistorischen, klassischen und frühchristlichen Altertümer, Swastika, in DACL III/2, 1914.

GCBI = Godišnjak, Centar za balkanološka ispitivanja, Annuaire, Centre d'études balkaniques, Sarajevo.

HAUSSING 1965 = H. W. Haussig (éd.), Götter und Mythen im Vorderen Orient, Bd. 1, Stuttgart 1965.

HZB = Hrvatski znanstveni zbornik (Kroatischer wissenschaftlicher Sammelband), Zagreb.

JNES = Journal of Near Eastern Studies, Chicago.

JÖAI = Jahreshefte des Österreichischen Archäologischen Institutes, Wien.

LÄ = Lexikon der Ägyptologie, Wiesbaden.

Lex. Myth = W. H. Roscher (éd.), Ausführliches Lexicon der griechischen und römischen Mythologie, Leipzig 1909-1915.

LIMC = Lexicon Iconographicum Mythologiae Classicae – VII. Zürich-München 1994.

LTL = Lexicon totius latinitatis. IV. Patavii 1890.

MDAI = Mitteilungen des k. k. Deutschen Archäologischen Instituts, Athenische Abteilungen, Athen.

OGIS = W. Dittenberger, Orientis Graeci Inscriptiones Selectae I – II., Leipzig 1903-1905.

RA = Revue Archéologique, Paris.

RE = Wissowa-Kroll-Mittelhaus (éds.), Realencyclopdie der Classischen Altertumwissenschaft. Stuttgart - München 1894-1980.

RFFZd = Radovi Filozofskog fakulteta u Zadru (Arbeiten Filosophische Fakultät Zadar), Zadar.

RIU = L. Barkóczi - A. Mócsy, Die Römische Inschriften Ungarns I. Savaria, Scarbancia und die Limes Strecke Ad Fexum, Arrabona, Budapest, 1972.

RSS = L. Balla-T. P. Buócz-Z. Kádár-A. Mócsy-T. Szentléleky, Die römischen Steindenkmäler von Savaria, Budapest 1971.

RT = Recueil des travaux relatifs à la philologie et à l'archéologie égyptiennes et assyriennes, Paris.

SIRIS = L. Vidman, Sylloge inscriptioneum religionis Isiacae et Sarapiacae, Berlin 1969.

SZ = Senjski zbornik (Senjer Jahrbuch), Senj.

VAHD = Vjesnik za arheologiju i historiju dalmatinsku (Bulletin d'archéologie et d'histoire dalmate), Split.

VHAD = Vjesnik Hrvatskoga arheološkoga društva ("Vjesnik" der Kroatischen archäologischen Gesellschaft in Agram), Zagreb.

ZDMG = Zeitschrift der Deutschen Morgenländischen Gesellschaft, Wiesbaden.

Livres cités

ABRAMIĆ 1912 = M. Abramić, Führer durch das k. k. Staatsmuseum in S. Donato in Zara, Wien 1912.

ABRAMIĆ 1924-25 = M. Abramić, Poklopac rimskog sarkofaga iz Salone (Couvercle d' un sarcophage trouvé à Salone), VAHD 47-48, 1924-25, 68-70.

ABRAMIĆ 1939 = M. Abramić, Ninska Venera (Venus von Nin), Glasnik Primorske Banovine 2, 1939, 200-202.

ABRAMIĆ 1940 = M. Abramić, Archäologische Forschung in Jugoslawien, Bericht über den VI. Internationalen Kongress für Archäologie, Berlin 1940, 166-180.

ADAM 1764 = R. Adam, Ruins of the Palace of the Emperor Diocletian at Spalato in Dalmatia, London 1764.

ABUBAKR 1937 = A. el M. J. Abubakr, Untersuchungen über die ägyptischen Kronen, Glückstadt-Hamburg-New York 1937.

ADRIANI 1935-39 = A. Adriani, Sanctuaire de l'époque romaine à Ras el-Soda, Annuaire du Musée Gréco-Romain 1935-39, 136-148.

ADRIANI 1940-1950 = A. Adriani, Municipalité d'Alexandrie, Trouvaille à Ras el-Soda, Annuaire du Musée Gréco-Romain, III, 1940-1950, 28-46.

ALDRED 1969 = C. Aldred, The "New Year" gifts to the Pharaoh, JEA 55, 1969, 73-81.

ALFANO 1989-1990 = C. Alfano, Le offerte alle divinità egizie a Roma in età imperiale, Scienze dell'Antichità 3-4, Roma 1989-1990, 785-791.

ALFANO 1990 = C. Alfano, Problemi archeologici relativi all' identificazione dei luoghi di culto egizio a Roma, in: Roma e l'Egitto nell'antichità classica, Roma 1990, 41-46.

ALFANO 2000 = C. Alfano, L' Iseo Campense in Roma: relazione preliminare sui nuovi ritrovamenti, manuscript received in 2000.

ALFÖLDI 1937 = A. Alföldi, Isis-szertartások Rómában a negyedik század keresztény császárai alatt (A festival of

Isis in Rome under the Christian emperors of the IVth century), Dissertationes Pannonicae ex Inst. Numism. et Arch. Univ. de Petro Pázmány nom. Bp. prov. ser. II. fasc 7. Budapest, 1937.

ALFÖLDI 1942 = A. Alföldi, Aquincum a későrómai világban, in: K. Szendy (éd.), Budapest Története, Budapest 1942, 670-745.

ALFÖLDI 1964-1965 = A. Alföldi, Die Alexandrinischen Götter und die vota publica am Jahresbeginn, Jahrbuch für Antike und Christentum 8-9, 1964-1965, 53-87.

ALFÖLDI 1982, 1990 = A. Alföldi - E. Alföldi, Die Kontorniat-Medaillons. Tl. 1. Katalog, in: Anzeiger f. d. Altertumswissen-schaft 35, Ht. 1/2 1, Berlin 1982, 98-115. Tl. 2. Text, in: Antike Münzen u. geschnittene Steine 6, Berlin-New York 1990.

ALFÖLDY 1963 = G. Alföldy, Die Geschichte des religiösen Lebens in Aquincum, Bud. Rég. 20, 1963, 47-69.

ALLEN 1985 = M. L. Allen, The Terracotta Figurines from Karanis: A Study of Technique, Style and Chronology in Fayoumic Coroplastics, Diss. Ann Arbor 1985.

ALLIOT 1949-1954 = M. Alliot, Le culte d'Horus à Edfou aux temps des Ptolémées. Fasc. 1-2, BiÉ 20, Le Caire 1949-1954.

ARSLAN 1997 = E. A. Arslan (éd.), Iside, il mito il misterio e la magia, Milano 1997.

ASSMANN 1979 = J. Assmann, Isis und die ägyptischen Mysterien, Aspekte der spätägyptischen Religion, Wiesbaden 1979, 93-115.

ASSMANN 1990 = J. Assmann, Ma'at. Gerechtigkeit und Unsterblichkeit im Alten Ägypten, München, 1990.

BALDACCI 1987 = T. Baldacci, Musica, danza e giochi, Civiltŕ degli Egizi, La vita quotidiana, Milano 1987, 232-247.

BALLA 1971a = L. Balla, Cultes orientaux à Savaria, in: J. Szabadfalvi-Z. Újvári (éds.), Studia ethnographica et folkloristica in honorem Béla Gunda, Debrecen 1971

BALLA 1971b - L. Balla, Religion und Religiöses Leben, in RSS 1971, 37-45.

BALLA 1974 = L. Balla, Á propos des périodes de l'Iséum de Savaria, in: Studia Aegyptiaca 1, Budapest 1974, 1-12.

BALLET 1988 = P. Ballet, Essai de recherches sur le culte d'Harpocrate: figurines en terre cuite d'Égypte et du Bassin Méditerranéen aux époques hellénistiques et romaines, Diss., Paris 1988.

BăRBULESCU 1984 = M. Bărbulescu, Interferențe spirituale în Dacia romană, Cluj-Napoca 1984.

BăRBULESCU 1994 = M. Bărbulescu, Potaissa. Studiu monografic, Dissertationes Musei Potaissensis I, Turda 1994.

BARKÓCZI 1964 = L. Barkóczi, The population of Pannonia from Marcus Aurelius to Diocletian, Acta Arch. XVI, 1964, 257-356.

BARTA 1968 = W. Barta, Das Götterkultbild als Mittelpunkt bei Prozessionsfesten, MDAIK 23, 1968, 75-78.

BAUS 1940 = K. Baus, Der Kranz in Antike und Christentum, Bonn 1940.

BAUVAL-GILBERT 1997 = R. Bauval – A. Gilbert, Il mistero di Orione, Milano 1997.

BAYER-NIEMEIER 1988 = E. Bayer-Niemeier, Liebighaus – Museum Alter Plastik. Bildwerke der Sammlung Kaufmann. Band I. Griechisch-römische Terrakotten, Melsungen 1988.

BECHER 1970 = I. Becher, Antike Heilgötter und die römische Staatsreligion. Philologus 114, 1970, 211-255.

BELLIDO 1949 = A. Garcia y Bellido, Esculturas romanas de Espańa y Portugal, Madrid 1949.

BELLIDO 1967 = A. Garcia y Bellido, Les religions orientales dans l' Espagne romaine, EPRO 5, Leiden 1967.

BENEDITE 1904 = G. Bénédite, Une nouvelle représentaion d' Horus légionaire, RA 1, Paris 1904, 111-118.

BERGMAN 1968 = J. Bergman, Ich bin Isis. Studien zum memphitischen Hintergrund der griechischen Aretalogien. Uppsala 1968.

Berlin 1990 = Pharaonen-Dämmerung. Wiedergeburt des alten Ägypten. Ausstellungskatalog, Berlin 1990.

BERSA 1906 = G. Bersa, Le lucerne fittili romane di Nona, sep. BASD 29, 1906, 76-86.

BERSA 1915 = G. Bersa, Le lucerne fittili romane di Nona conservate al Museo Archeologico in San Donato di Zara, BASD 38, 1915, 46-86.

BESQUE 1992 = S. Besque, Musée du Louvre. Catalogue raisonné des Figurines et reliefs en terres-cuite grecs, étrusques et romains, IV, II: Époques hellénistique et romaine. Cyrénaïque, Égypte Ptolémaïque et Romaine, Afrique du Nord et Proche-Orient, Paris 1992.

BIANCHI 1980 = U. Bianchi, Iside dea misterica, in: Quando, Perennitas: Studi di A. Brelich, Roma 1980, 9-36.

BIEBER 1961 = M. Bieber, The Sculpture of the Hellenistic Age, New York 1961.

BIETAK 1968 = M. Bietak, Vorläufiger Bericht über die erste und zweite Kampagne der österreichischen Ausgrabungen auf Tell Ed-Dab'a im Ostdelta Ägyptens (1966, 1967), MDAIK 23, 1968.

BIRLEY 1966 = A. Birley, Marcus Aurelius, London 1966.

BISSING 1925 = F. W. von Bissing, Eine hellenistische Bronzefigure des Gottes Bes, MDAI 50, 1925, 123-132.

BLECH 1982 = M. Blech, Studien zum Kranz bei den Griechen, Berlin 1982.

BLINKENBERG 1904 = C. Blinkenberg, Archäologische Studien. Darstellungen des Sabazios und Denkmäler seines Kultes, Copenhague-Leipzig 1904.

BORHY 1997 = L. Borhy, Glasgemme mit antoninischem Frauenporträt aus Brigetio (Komárom/Szôny, Ungarn), AA 1997, 97-100.

BRECCIA 1930 = Ev. Breccia, Terrecotte figurate greche e greco-egizie del Museo di Alessandria, Monuments de l'Égypte gréco-romaine II, 1, Bergamo 1930.

BRECCIA 1934 = Ev. Breccia, Terrecotte figurate greche e greco-egizie del Museo di Alessandria, Monuments de l'Égypte gréco-romaine II, 2, Bergamo 1934.

BRELICH 1938 = A. Brelich, Aquincum vallásos élete, Laureae Aquincensis 1, Diss. Pann. ser. 2, no. 10, Budapest 1938, 20-142.

BRESCIANI 1968 = E. Bresciani, Rapporto Preliminare delle

campagne di scavo 1966 e 1967 (Missione di scavo a Medinet Madi, Fayum, Egitto), Milano 1968.
BRESCIANI 1976 = E. Bresciani, Rapporto preliminare delle campagne di Scavo 1968 e 1969, Milano 1976.
BRESCIANI 1997 = E. Bresciani, La Iside di Medinet Madi, in: ARSLAN 1997, 37-41.
BRUNEAU 1970 = Ph. Bruneau, Recherches sur les cultes de Délos à l'époque hellénistique et à l'époque impériale, Paris 1970.
BRUNŠMID 1898 = J. Brunšmid, Arheološke bilješke iz Dalmacije i Panonije (II), VHAD n.s. 3, 1898, 150-205.
BRUNŠMID 1903-04 = J. Brunšmid, Kameni spomenici hrvatskoga narodnoga muzeja u Zagrebu (Die Steindenkmäler des kroatischen Nationalmuseums in Zagrcb), VHAD 7, 1903-04, 209-240.
BRUNŠMID 1914 = J. Brunšmid, Antikni figuralni brončani predmeti, VHAD 13, 1914, 207-268.
BRUSIĆ 2000 = Z. Brusić, Vinske amfore i posuđe za posluživanje vina sa hrvatske obale Jadrana (Wine Amphorae and wine seving vessels from the croatian coast of the Adriatic Sea), Annales. Anali za istrske in mediter-anske študije 22, 2000, 531-542.
BUDISCHOVSKY 1973 = M. C. Budischovsky, Jupiter-Ammon et Meduse dans les forum du nord de l'Adriatique, Aquileia Nostra XLIV, Aquileia 1973, 201-220.
BUDISCHOVSKY 1977 = M. C. Budischovsky, La diffusion des cultes isiaques autour de la mer Adriatique. I. Inscriptions et monuments, Leiden 1977.
BULIĆ 1886 = F. Bulić, Descrizione delle lucerne fittili che si conservano nell' I. R. Museo di Spalato, BASD 9, 1886, 21-24.
BULIĆ 1887 = F. Bulić, Descrizione delle lucerne fittili che furono aquistate dell' I. R. Museo in Spalato nella 1887, BASD 10, 1887, 31-33, 46-48.
BULIĆ 1908 = F. Bulić, Materiale e provenienza della pietra, delle colonne nonché delle sfingi del Palazzo di Diocleziano a Spalato, BASD 31, 1908, 86-127.

BULIĆ 1909 = F. Bulić, Ritrovamenti antichi a Oklaj di Promina, BASD 32, 1909, 45-48.

BULIĆ 1910 = F. Bulić, Escavi nella necropoli antica pagana di Salona detta Hortus Metrodori negli anni 1909 e 1910, BASD 33, 1910, 3-66, 130-135.

BULIĆ 1926-27 = F. Bulić, Riječ dvije o figuri Besa" na jednoj brončanoj vazi u splitskom Arheološkom muzeju, VAHD 49, 1926-27, 105.

BULIĆ-KARAMAN 1967 = F. Bulić – Lj. Karaman, Palača cara Dioklecijana u Splitu, Zagreb 1927.

BUÓCZ 1964 = T. P. Buócz, A szombathelyi Savaria Múzeum kőtára 9. Savaria Múzeum Közleményei 30, 1964, 263-279.

BUÓCZ 1968 = T.P. Buócz, Savaria topográfiája, Szombathely 1968.

BUÓCZ 1978 = T.P. Buócz, Probleme der Stadtstruktur von Savaria, Forschungsber. z. Ur- u. Frühgesch. 10, 1978, 887-88.

BURKERT 1991 = W. Burkert, Antichi culti misterici, Roma-Bari 1991.

BURR-THOMPSON 1934 = Do. Burr-Thompson, Terracottas from Myrina in the Museum of Fine Arts, Boston, Wien 1934.

BUŠKARIOL 1988 = F. Buškariol, Zbirka Machiedo u Arheološkom muzej u Splitu, Prilozi povijesti otoka Hvara, Hvar 1988, 31-41.

CAMBI 1963-65 = N. Cambi, Nove potvrde egipatskih kultova u antičkoj provinciji Dalmaciji (Nouvelles confirmations des cultes égyptiens dans l' antique province de Dalmatie), VAHD 65-67, 1963-65, 85-112.

CAMBI 1980 = N. Cambi, Jupiter Amon na dva spomenika iz Arheološkog muzeja u Splitu (Jupiter Amon on two monuments in the Archaeological Museum in Split), Gunjačin zbornik, Zagreb 1980, 43-50.

CAMBI 1980a = N. Cambi, Enonska Venera Anzotika (Venus Anzotica d' Aenona), Diadora 9, 1980, 273-288.

CAMBI 1993 = N. Cambi, Stele iz kasnoantičke grobnice u Dugopolju (Die Stelen der spätantiken Grabstätte in Dugopolje), VAHD 86, 1993, 147-181.

CAMBI 2000 = N. Cambi, Imago animi. Antički portret u Hrvatskoj (Imago animi. Ancient Portraits in Croatia), Split 2000.

CAPART 1932 = J. Capart, Le naos de décret de Rosette, in: Studies presented to F. L. Griffith, London 1932, 277-278.

CASTIGLIONE 1978 = L. Castiglione, Nouvelles données archéologiques concernant la génése du culte de Sarapis, in: M.A. de Boer - T.A. Edridge (éds.), Hommages à Maarten J. Vermaseren I. EPRO 68. Leiden, 1978, 208-232.

CAUVILLE 1980 = S. Cauville, Une offrande spécifique d'Osiris: le récipient de dattes, RdÉ 32, 1980, 47-64.

CHAMAY 1996 = J. Chamay, Bronze et Or. Visages de Marc Aurèle, Musée d'Art et Histoire, Genève 1996.

CHANDLER – FOX 1974 = Tertius CHANDLER - Gerard FOX, Three Thousand Years of Urban Growth, New York 1974.

CLÉDAT 1912 = Jean Clédat, Fouilles à Qasr-Gheit (Mai 1911), ASAE 12, 1912, 145-168.

CLERC – LECLANT 1994 = G. Clerc - J. Leclant, Sarapis. in: LIMC VII/1, 1994, 666-692.

COLNAGO – KEIL 1905 = A. Colnago - J. Keil, Archäologische Untersuchungen in Norddalmatien, JÖAI 8, 1905, 31-60.

CONZE 1877 = A. Conze, Sphinx Amenophis III, AEM I, 1877, 95-97.

CONZE 1878 = A. Conze, La sfinge di Amenhotep III in Spalato, BASD 1, 1878, 43-45.

CUMONT 1959 = F. Cumont, After Life in Roman Paganism, New York 1959.

COONEY 1972 = J.D. Cooney, Harpocrates, The Dutiful Son, Bulletin of the Cleveland Museum of Art 59, 1972, 284-290.

DEGRASSI 1937 = A. Degrassi, I magistri Mercuriales di Lucca e la dea Anzotica di Aenona, Athenaeum 15, 1937, 284-288.

DEGRASSI 1962 = A. Degrassi, Salvore – scoperta d'antichità romane, Scritti vari di antichità, II, Roma 1962, 881-886.

DEMISCH 1977 = H. Demisch, Die Sphinx, Geschichte ihrer Darstellung von den Anfängen bis zur Gegenwart, Stuttgart 1977.

DÉRI 1922 = F. Déri, A Debreceni Déri Museum gyűjteményeinek leírása (Description des collections du Musée Déri à Debrecen), Debreczen 1922.

DERCHAIN-HUBAUX 1953 = P. Derchain - J. Hubaux, Vespasien au Sérapéum, Latomus 12, 1953, 38-52.

DESSENE 1957 = A. Dessene, Le Sphinx. Étude Iconographique. I. Des origines à la fin du second millénaire, Paris 1957.

DIEZ 1986 = E. Diez, Die thebanischen Sphinx in Pannonien und Noricum, in: E. Böhr und W. Martini (éds.), Studien zur Mythologie und Vasenmalerei. Festschrift für K. Schauenburg, Mainz 1986, 247-252

DOBROVITS 1937 = A. Dobrovits, Harpokrates, in: Jubilee Volume in honour of E. Mahler, Budapest 1937. 72-122.

DOBROVITS 1934 = A. Dobrovits, Egyiptom és hellenizmus (Ägypten und der Hellenismus), Budapest 1934.

DOBROVOTS 1939 = A. Dobrovits, Az Egyiptomi Gyűjtemény (La collection égyptienne), Budapest, 1939.

DOBROVITS 1943 = A. Dobrovits, Az egyiptomi kultuszok emlékei Aquincumban (Les monuments des cultes égyptiens à Aquincum), Bud. Rég. XIII, 1943, 47-75.

DOBROVITS 1954 = A. Dobrovits, Egyiptomi amulett szarmata sírból. A magyar és orosz iparművészet történeti kapcsolatai (Amulette égyptienne d'un tombeau sarmatique. Les rapports historiques entre les arts décoratifs hongrois et russe), Budapest 1954, 9-34.

DONADONI 1997 = S. F. Donadoni, Iside faraonica, in: ARSLAN 1997, 32-36.

DONATELLI 1988 = L. Donatelli, Piccoli oggetti di corredo fune-rario nelle varie epoche, Civiltà degli Egizi. Le credenze religiose, Torino 1988, 198-211.

DÖLGER 1936 = Fr. J. Dölger, Anulus pronubus. Antike und Christentum 5, 1936, 188-200.

DREXEL 1909 = Fr. Drexel, Alexandrinische Silbergefässe der Kaiserzeit, BJ 118, 1909, 176-235.

DREXLER 1900 = W. Drexler, Il culto delle divinità egiziane in Dalmazia, BASD 23, 1900, 145-158.

DUNAND 1973 = F. Dunand, Le culte d' Isis dans le basin oriental de la Méditerranée, I-III, EPRO 26, Leiden 1973.

DUNAND 1984 = Fr. Dunand, Religion populaire et iconographie en Égypte hellénistique et romaine, in: Visible Religion. Annual for Religious Iconography III. Popular Religion, Leyde 1984.

DUNAND 1979 = Fr. Dunand, Religion populaire en Égypte Romaine. Les terres cuites isiaques du Musée du Caire, EPRO 76, Leiden 1979.

DUNAND 1990 = Fr. Dunand, Catalogue des terres cuites gréco-romaines d'Égypte, Musée du Louvre, Département des Antiquités Égyptiennes, Paris 1990.

DUNAND 1991 = Fr. Dunand - Ch. Zivie-Coche, Dieux et hommes en Égypte 3000 av. J.-C. - 395 apr. J.-C., Paris 1991.

EDGAR 1904 = C.C. Edgar, Catalogue général des antiquités égyptiennes du Musée du Caire. Greek Bronzes, Le Caire 1904.

ENGELBACH 1915 = R. Engelbach, Riqqeh and Memphis VI, ERAP 1913, London 1915.

ERMAN 1934 = A. Erman, Die Religion der Ägypter, Berlin-Leipzig 1934.

ERMAN 1952 = A. Erman, La religion égyptienne, Paris 1952.

ERNST 2001 = H. Ernst, Altar oder Barkenuntersatz, GM 180, 2001, 57-66.

EWIGLEBEN 1991 = C. Ewigleben – J. von Grumbkov (éds.), Götter, Gräber und Grotesken. Tonfiguren aus dem Alltagsleben im römischen Ägypten. Museum für Kunst und Gewerbe Hamburg, 1991.

FACCIOLATI-FORCELLINI-FURLAMENETTI 1890 = J. Faccio-lati - Aeg. Forcellini- J. Furlanetti, Sphinx, in: LTL IV., p. 448.

FAULKNER 1925 = R. O. Faulkner, The King and the Star-Religion in the Pyramid Texts, JNES 25, 1925, 153-161.

FISCHER 1994 = J. Fischer, Griechisch-römische Terrakotten aus Ägypten. Die Sammlungen Sieglin und Schreiber. Dresden - Leipzig - Stuttgart - Tübingen. Tübinger Studien zur Archäologie und Kunstgeschichte. Bd. 14, Tübingen 1994.

FITTISCHEN = ZANKER 1985 = K. Fittschen - P. Zanker, Katalog der römischen Porträts in den Capitolinischen Museen und den anderen kommunalen Sammlungen der Stadt Rom. I.: Kaiser- und Prinzenbildnisse, Mayence 1985.

FITZ 1972 = J. Fitz, Les Syriens à Intercisa, Collection Latomus 122, Bruxelles 1972.

FITZ 1978 = J. Fitz, Grenzberichtung im Jahre 214 zwischen Pannonia Superior und Inferior, Alba Regia 16, 1978, 71-86.

FITZ 1992-1994 = J. Fitz, Die Verwaltung Pannoniens in der Römerzeit I-IV., Budapest, 1992-1994.

FITZ 1998 = J. Fitz (éd.), Religions and Cults in Pannonia. Székesfehérvár 1998.

FOURLAS 1971 = A. A. Fourlas, Der Ring in der Antike und im Christentum. Der Ring als Herrschaftssymbol und Würdezeichen, Münster 1971.

FRANKFURTER 1998 = D. Frankfurter, Religion in Roman Egypt, Princeton 1998.

FRASER 1972 = P.M. Fraser, Ptolemaic Alexandria, Oxford, 1972.

FRASER 1967 = P. M. Fraser, Current Problems concerning the Early History of the Cult of Serapis, in: Opuscula Atheniensia 7, Athens 1967, 23-45.

FROEHNER 1867 = W. Froehner, Terres cuites d'el Kantara, Bull. Archéol. Du Musée Parent, 1. Oct, 1867.

GABLER 1990 = D. Gabler, Kereskedelem, in: A. Mócsy - J. Fitz (éds.), Pannonia Régészeti Kézikönyve, Budapest 1990.

GALLI 1939 = E. Galli, Aenona odierna Nin – Illyricum: Santuario di Venere "Anzotica", Athenaeum 17, 1939, 50-53.

GALLO 1997 = P. Gallo, Luoghi di culto e santuari isiaci in Italia, in: ARSLAN 1997, 290-383.

GANSCHNITZ in: RE II/1 = R. Ganschinitz, Ringe in Folklore. in: RE II/1. 833-841.

GANZERT-KOCKEL 1988 = J. Ganzert - V. Kockel, Augustus-forum und Mars-Ultor-Tempel, in: Kaiser Augustus und die verlorene Republik, Berlin 1988, 149-199.

GAUKNER – MERLIN 1910 = M. P.Gaukner - M.A. Merlin, Catalogue du Musée Alaoui. Supplément, Paris 1910.

GENAILLE 1976-1977 = N. Genaille, Le sistre Strozzi (à propos des objets cultuels isiaques en Italie), BSFE 77-78, 1976-1977, 56-67.

GENAILLE 1994 = N. Genaille, Instruments du culte isiaques fi-gurés sur trois monuments funeraires de Rome, in: C. Berger-G. Clerc- N. Grimal (éds.), Hommages à Jean Leclant, BdE, III, Paris 1994, 223-234.

GESZTELYI 1998 = T. Gesztelyi, Pannoniai vésett ékkövek, M 2, Budapest 1998.

GESZTELYI 2000 = T. Gesztelyi, Antike Gemmen im Ungarischen Nationalmuseum. Catalogi Musei Nationalis Hungarici. Ser. Arch. III., Budapest 2000.

GIRON 1924 = N. A. Giron, Glanures de mythologie syro-égyptienne, BIFAO XXIII, 1924, 11-25.

GIUNIO 1997 = K. A. Giunio, Tipologija rimskih hramova carskog doba i njihov odraz na istočnoj obali Jadrana (Typology of the Roman Imperial Temples and their reflection on the Eastern Adriatic Coast), Zagreb 1997.

GIUNIO 1999 = K. A. Giunio Neke bilješke o zadarskom forumu i kapitoliju (Notes on the Forum and the Capitolium in Zadar), Histria Antiqua 5, Pula 1999, 55-66.

GIUNIO 2000, K. A. Giunio, Egipatski amulet u Zadru (Egyptian Amulet in Zadar), Diadora 20, Zadar 2000, 177-191.

GLAVIČIĆ 1981-82 = A. Glavičić, Arheološki nalazi iz Senja i okolice (V.), SZ 9, Senj 1981-82, 63-77.

GLAVIČIĆ 1993 = M. Glavičić, Prilozi proučavanju poleogeneze Senije (Contributions towards the study of the Palaeogenesis and Urban development of antique Senia), RFFZd 32/9, 1993, 79-104.

GLAVIČIĆ 1994 = M. Glavičić, Natpisi antičke Senije (The inscriptions from Roman Senia), RFFZd, 33/10, Zadar 1994, 55-82.

GNIRS 1915 = A. Gnirs, Pola, ein Führer durch die antiken Baudenkmäler und Sammlungen, Wien 1915.
GONZENBACH 1957 = V. von Gonzenbach, Untersuchungen zu den Knabenweihen im Isiskult der Römischen Kaiserzeit, Bonn 1957.
GOYON 1972 = J. C. Goyon, Confirmation du pouvoir royal lors du Nouvel An (Brooklyn Museum Papyrus 47.218.59), BiÉ 52, Le Caire 1972.
GRAINDOR 1939 = P. Graindor, Terres cuites de l'Égypte gréco-romaine, Antwerpen 1939.
GREIFENHAGEN 1961 = A. Greifenhagen, Goldschmuck-Verluste, Antiquarium Berlin 1946-1947, AA 1961, 82-130.
GRENIER 1977 = J.-C. Grenier, Anubis alexandrin et romain, EPRO 57, Leiden 1977.
GRIFFITHS 1980 = J. Gwyn Griffiths, The Origins of Osiris and His Cult, Studies in the history of Religion, Leiden 1980.
GRIFFITH 1975 = J. G. Griffiths, Aus dem ersten Hymnus Isidors 26 (SEG 8,548). The Isis-Book, Leiden 1975.
GRIMM 1969 = G. Grimm, Die Zeugnisse Ägyptischer Religion und Kunstelemente im römischen Deutschland, EPRO 12, Leiden 1969.
GRIMM 1997 = A. Grimm, Iside imperiale. Aspetti storico-cultu-rali del culto isiaco al tempo degli imperatori romani, in: ARSLAN 1997, 120-133.
GRIMM - KESSLER - MEYER 1994 = A. Grimm - D. Kessler - H. Meyer, Das Obelisk des Antinoos, München 1994.
GUÉRAUD-KUENTZ 1936 = O. Guéraud -Ch. Kuentz, Le crépuscule d'un dieu. Harpocrate Ξερνιβοπαστης, ASAE 36, 1936, 115-123.
GUTCH 1898-1899 = C. Gutch, Excavations at Naucratis. D. The Terracottas, ABSA V, 1898-1899, 67-97.
GYŐRY 1983 = H. Győry, Harpocrates körmeneteinek ábrázolása a terrakottákon, Studia Iuvenum in Honorem V. Wessetzky, Egyiptológiai Füzetek I., Budapest 1983.
GYŐRY 1997 = H. Győry, Az első magyar ásatás Egyiptomban. Válogatás a Gamhudi ásatás leleteiből, 1997. (La première fouille hongroise en Égypte. Choix des trou-

vailles de l'expédition à Gamhoud), Budapest 1997.
HAJNÓCZI – SZENTLÉLEKY 1964 = Gy. Hajnóczy – T. Szentléleky, Római kori homlokzat helyreállítása Szombathelyen. Magyar Műemlékvédelem 2, 1959-60. Budapest, 1964.
HALSBERGHE 1972 = G. H. Halsberghe, The cult of Sol Invictus, Leiden 1972.
HANNIG 1995 = L.R. Hannig, Hannig Lexica 1. Die Sprache der Pharaonen. Grosses Handwörterbuch, Ägyptisch-Deutsch, Mainz 1995.
HARDER 1944 = R. Harder, Karpocrates von Chalkis und die memphitische Isis Propaganda, Berlin 1944.
HARRIS 1965 = E. and J. R. Harris, The Oriental Cults in Roman Britain, Leiden 1965.
HAUSMANN 1984 = Ulrich Hausmann, Zur Eroten- und Gallier - Ikonographie in der alexandrinischen Kunst in: N. Bonacasa - A. di Vita (éds.), Studi e materiali – Istituto di Archeologia Università di Palermo 5., Alessandria e il mondo ellenistico - romano. Studi in onore di Achille Adriani, Roma 1984.
HAUSSIG 1965 = H.W. Haussig (Hrg.), Götter und Mythen im Vorderen Orient, Bd.1, Stuttgart 1965.
HENNE 1925 = H. Henne, Rapport sur les fouilles de Tell Edfou, FIFAO II, 1923-1924, Le Caire 1925.
HERBIG 1929 = R. Herbig, Sphinx, in: RE II/6/3. 1703-1749.
HERMITAGE 1976 = Antyicsnaja Koroplasztika, Leningrad – Hermitage 1976.
HEGEDUS 1998 = T. Hegedus, The Urban Expansion of the Isis Cult: A Quantitative approach, in: Studies in Religion/Sciences Religieuses 27/2, 1998, 161-178.
HEKLER 1909 = A. Hekler, Die hellenistischen Bronzegefässe von Egyed, JdAI 24, 1909, 36.
HENKEL 1913 = F. Henkel, Die römischen Fingerringe der Rheinlande und der benachbarten Gebiete, Berlin 1913.
HEYOB 1975 = Sh. K. Heyob, The Cult of Isis among Women in the Graeco-Roman World, EPRO 51, Leiden 1975.
HELM 1956 = R. Helm (éd.), Apuleius, Metamorphosen, Berlin 1956.

HILL 1946 = D. K. Hill, Material on the Cult of Sarapis, Hesperia 15, 1946, 60-72.

HIMMELMANN 1983 = N. Himmelmann, Alexandria und der Realismus in der griechischen Kunst, Tübingen 1983.

HOFFMAN 1993 = P. Hoffman, Der Isis-Tempel in Pompeji, Münster-Hamburg 1993.

HORNBOSTEL 1973 = W. Hornbostel, Sarapis. Studien zur Überlieferungsgeschichte den Erscheinungsformen und Wandlun-gen der Gestalt eines Gottes, EPRO 32, Leiden 1973.

HORNBOSTEL etc. 1973 = W. Hornbostel etc., Kunst und Altertum am Rhein. Führer des Rheinischen Landesmuseums in Bonn, Nr. 48. Antiken aus Rheinischem Privatbesitz, Köln 1973.

HORNBOSTEL etc. 1977 = W. Hornbostel etc., Kunst der Antike. Schätze aus Norddeutschen Privatbesitz. Museum für Kunst und Gewerbe, Hamburg-Mainz 1977.

HORNBOSTEL 1978 = W. Hornbostel, Sarapiaca I., in: M.A. de Boer - T.A. Edridge (éds.), Hommages à Maarten J. Verma-seren I. EPRO 68. Leiden, 1978.

HORVÁTH 1985 = J. Horváth, A magyar egyiptológia története (Histoire de l'égyptologie hongroise), Egyipotológiai Füzetek II., Budapest 1985.

HÖFLER 1935 = A. Höfler, Der Sarapishymnus des Ailios Aristeides, Diss. Tübingen, Tübingen 1933 (1935).

HÖLBL 1983 = G. Hölbl, Ägyptische Religion im Römischen Reich, in: Spätantike und frühes Christentum, Ausstellung im Liebighaus, Museum alter Plastik, Frankfurt am Main, 16. Dezember 1983 bis 11. März 1984, Frankfurt am Main 1983. 98-110.

HÖLBL 1984 = G. Hölbl, Sarapis, in: LÄ V. 1984, 870-874.

ILBERG 1909-1915 = J. Ilberg, Sphinx, in: Lex. Myth. IV., 1298-1508.

IVERSEN 1968 = E. Iversen, Obelisks in Exile I., Copenhagen 1968, 80-92.

JÉQUIER 1910 = C. Jéquier, La sfinge nel Peristilio del Palazzo di Diocleziano, BASD 33, 1910, 174-179.

JOBST 1992 = H. Jobst, Ägyptische Kulte, in: W. Jobst (éd.), Carnuntum, Das Erbe Roms an der Donau, Wien 1992.

JURKIĆ-GIRARDI 1970 = V. Jurkić Girardi, Meduze na reljefima Arheološkog muzeja Istre u Puli (Medusa auf den Reliefen des Archäologischen Museums Istriens in Pula), Histria Archaeologica 1/2, Pula 1970, 27-52.

JURKIĆ-GIRARDI 1972 = V. Jurkić-Girardi, Rasprostranjenost kulta Magnae Matris na području Istre u rimsko doba, Histria Archaeologica 3/1, Pula 1972, 41-76.

JURKIĆ-GIRARDI 1976 = V. Jurkić-Girardi, Izbor antičke kultne plastike na području Istre (Die Auswahl der antiken Kultplastik auf dem Gebiet Istriens), Materijali 12, Arheološki problemi na jugoslavenskoj obali Jadrana (Problèmes archéologiques de la cőte yougoslave de l'Adriatique), Zadar 1976, 209-223.

JURKIĆ-GIRARDI – DŽIN - URANIĆ 2001 = V. Jurkić-Girardi – K. Džin – I. Uranić, Egipatska religija i antička Istra (Egyptian Religion and ancient Istria). Catalogue, Pula 2001.

KABA 1991 = M. Kaba, Thermae maioris legionis II. Adiutricis, Budapest 1991.

KÁKOSY 1967 = L. Kákosy, Augustus és Egyiptom, AT 14, 1967, 307-315.

KÁKOSY 1983 = L. Kákosy, Die Kronen im Spätägyptischen Totenglauben, in: Das römisch-byzantinische Ägypten, Acten des Symposiums 26-30. September 1978 in Trier, Aegyptiaca Treverensia 2, Mainz 1983.

KÁKOSY 1997 = L. Kákosy, Riti iniziatici e misteri nel culto isiaco, In: ARSLAN 1997, 148-150.

KÁKOSY 1999 = L. Kákosy, Mysteries in the Isiac Religion, Acta Ant. 39, 1999, 159-164.

KAMAL 1905 = A. Bey Kamal, Stèles Ptolémaïques et romaines, Catalogue Général du Musée du Caire, CG 22001-22208, Le Caire 1905.

KAMAL 1908 = Ahmed Bey Kamal, ASAE 9, 1908, 8-30.

KARAGORGHIS 1959 = V. Karageorghis, Chronique des fouilles et découvertes archéologiques à Chypre en 1958,

BCH 83, 1959, 336-361.

KAPLONY 1966 = P. Kaplony, Das Vorbild des Königs unter Sesostris III., Orientalia 35, 1966, 403-412.

KARAMAN 1937 = Lj. Karaman, Staroegipatska sfinga otkopana u Splitu, Jugoslavenski istoriski časopis III/1-4, Beograd 1937, 615-616.

KARO 1901 = G. Karo, Die griechisch-römischen Altertürmer im Museum zu Kairo, AA 1901, 199-213.

KATER-SIBBES 1973 = G.J.F. Kater-Sibbes, Preliminary Catalogue of Sarapis Monuments, EPRO 36, Leiden 1973.

KAUFMANN 1913 = C.M. Kaufmann, Ägyptische Terrakotten der griechisch-römischen und koptischen Epoche vorzugsweise aus der Oase el Faijum, Kairo 1913.

KAUFMANN 1915 = Graeco-ägyptische Koroplastik. Terrakotten der griechisch-römischen und koptischen Epoche aus der Faijum-Oase und anderen Fundstätten, Kairo 1915.

KEES 1915 = Kees, Eine Liste memphitischer Götter im Tempel von Abydos, RT 37, 1915, 57-76.

KERÉNYI 1977 = K. Kerényi, Görög mitológia (Die Mythologie der Griechen), Budapest 1977.

KLEINER 1984 = G. Kleiner, Tanagrafiguren. Untersuchungen zur hellenistischen Kunst und Geschichte, Berlin-New York 1984 (reprint).

KLEMENC 1940 = J. Klemenc, Senj u prethistorijsko i rimsko doba, in: Senj, Hrvatski kulturni spomenici 1, Zagreb 1940.

KOVÁCS 1984 = V. Kovács, Bronzes romains figurés et appliqués et leurs problèmes techniques, Actes du VIIe Colloque international sur les Bronzes Antiques, Székesfehérvár 1984.

KRAUSKOPF 1994 = I. Krauskopf, Oidipous, in: LIMC VII/1 (Oidipous-Theseus), 1-15 et VII/2: illustrations.

KREIKENBOM 1992 = D. Kreikenbom, Griechische und römische Kolossalportrait bis zum späten ersten Jahrhundert nach Christus, JdI Ergänzungsheft 27, Berlin-New York 1992.

KOLB 1977 = F. Kolb, Zur Statussymbolik im antiken Rom,

Chiron 7, 1977, 239-259.
KRAUTHEIMER 1987 = R. Krautheimer, Rom. Schicksal einer Stadt 312-1308, München 1987.
KUBITSCHEK-LOEWY 1879 = W. Kubitschek - E. Loewy, Bericht über eine Reise in Ungarn, Slavonien und Croatien, AEM 3, 1879, 152-174.
KUNTIĆ-MAKVIĆ 1982 = B. Kuntić-Makvić, Žrtvenik iz Varvarije posvećen Izidi (L' autel de Varvaria, consacré à la déesse Isis), ARR 8-9, 1982, 151-157.
KUZMOVÁ 1997 = K. Kuzmová, Zur Interpretation einer Kopfdarstellung auf einer römischen Prunkfibel aus Iza, in: Roman Portraits. Artistic and Literary, Mainz 1997.
KUZSINSZKY 1897 = B. Kuzsinszky, Az Aquincumi Múzeum és kőemlékei, Bud. Rég. 5, 1897, 9-83.
KUZSINSZKY 1932 = B. Kuzsinszky, Az Aquincumi Múzeum és kőemlékei, Bud. Rég. 11, 1932, 143-157.
KYRIELEIS 1975 = H. Kyrieleis, Bildnisse der Ptolemäer, AF 2, Berlin 1975.
LABROUSSE 1952 = M. Labrousse, Bague à l'effigie de Sérapis-Ammon trouvée près de Royan (Charente-Maritime), RA II, 1952, 93-95.
LAKATOS 1961 = P. Lakatos, Beiträge zur Verbreitung der ägyptischen Kulte in Pannonien, Szeged 1961.
LANZA 1856 = F. Lanza, Monumenti inediti, Split 1856.
LAUMONIER 1921 = A. Laumonier, Catalogue de terres cuites du Musée Archéologique de Madrid. Bibliothèque de l'École de la Hautes Études Hispaniques. Fasc. II, Bordeaux - Lyon etc. 1921.
LECLANT 1971 = J. Leclant, Fouilles et travaux en Égypte et au Soudan 1969-1970. Orientalia NS 40, 1971, 224-266.
LECLANT 1976 = J. Leclant, Iconographie des petites bronzes d'Isis allaitant Horus exhumes en France, Annales de l'Université de Lyon, III, Lyon 1976. (1977).
LECLANT 1994 = C. Berger - G. Clerc - N. Grimal (éds.), Hommage ŕ J. Leclant, I-III., BiE 106/1-3, Le Caire 1994.
LECLANT 1997 = J. Leclant, Prefazione, in: ARSLAN 1997, 19-28.

LEHMANN-HAUPT in Lex. Myth = C.F. Lehmann-Haupt, Sarapis. in: Lex . Myth. 1909-1915, 338-382.

LEMBKE 1994 = K. Lembke, Das Iseum Campanense in Rom, Archäologie und Geschichte, 3, Heidelberg 1994.

LIPPOLD 1956 = G. Lippold, Die Sculpturen des Vatikanischen Museums, III.1, Berlin 1956.

LISE 1988 = G.Lise, Amuleti egizi, Milano 1988.

LUCIUS 1749 = J. Lucius (Lucić), Inscriptiones Dalmaticae..., Venezia 1673, 13, no. 8.

MAFFEI 1749 = S. Maffei, Museum Veronense, Verona 1749.

MAHLER 1913 = E. Mahler, Beöthy Zsolt egyiptológiai gyűjteménye (La collection égyptologique de Zsolt Beöthy), Budapest 1913.

MALAISE 1972 = M. Malise, Inventaire préliminaire des documents égyptiens découverts en Italie, ÉPRO 21, Leiden 1972.

MALAISE 1972a = M. Malaise, Les conditions de la pénétration et de diffusion des cultes égyptiens en Italie, ÉPRO 22, Leiden 1972.

MALAISE 1978 = M. Malaise, Documents nouveaux et points de vue récents sur les cultes isiaques en Italie, in: M.A. de Boer - T.A. Edridge (éds.), Hommages à Maarten J. Vermaseren I. EPRO 68. Leiden, 1978, 627-717.

MALAISE 1984 = M. Malaise, La diffusion des cultes égyptiens dans les provinces européennes de l'Empire Romain, in: ANRW II.17.3, Berlin - New-York 1984, 1615-1691.

MALAISE 1991 = M. Malaise, Harpocrate au pot, in: U. Verhoeven-E. Graefe (éds.), Religion und Philosophie im Alten Ägypten, Louvain 1991, 219-232.

MALAISE 1991a = M. Malaise, Harpocrate, la tortue et le chien. Contribution à l'iconographie du fils d'Isis, BSFE 122, 1991, 13-35.

MALAISE 1993 = M. Malaise, Le faucon et la chouette d'Harpocrate, in: Chr. CANNUYER - J-M. Kruchten (éds.), Individu, société et spiritualité dans l'Égypte pharaonique et copte, Mélanges Théodorides, Athen-Bruxelles-Mons 1993, 147-158.

MALAISE 1994 = M. Malaise, Questions d'iconographie harpocratique soulevées par des terres cuites d'Égypte grécoromaine, in: LECLANT 1994, 373-384.

MARASOVIĆ-MARASOVIĆ 1970 = J. and T. Marasović, Diocletian Palace, Zagreb 1970.

MARQUART 1879 = J. Marquart, Das Privatleben der Römer. Handbuch der römischen Alterthümer 7, Leipzig 1879.

MARSHALL 1907 = F.M. Marshall, Catalogue of the Finger Rings, Greek, Etruscan, and Roman in the Departments of Antiquities, British Museum, London 1907.

MARSHALL in: RE II/1 = F.M. Marshall, Ring. in: RE II/1 1914, 807-833.

MATIJAŠIĆ 1988 = R. Matijašić, Ageri antičkih kolonija Pula i Parentium (i njihova naseljenost od I. do III. stoljeća), Zagreb 1988.

MEDINI 1980 = J. Medini, Sabazijev kult u rimskoj provinciji Dalmaciji (Sabazius' Cult in the Roman Province of Dalmatia), VAHD 74, 1980, 67-88.

MELUCCOVACCARO 1989 = A. Meluccovaccaro (éd.), Marco Aurelio. Storie di un monumento e del suo restauro, Milan 1989.

MENSCHING o.J. = G. Mensching, Die Religion. Erscheinungs-formen, Strukturtypen und Lebensgesetze, München o. J.

MENZEL 1964 = H. Menzel, Römische Bronzen. Bildkataloge des Kestner-Museums Hannover VI., Hannover 1964.

MENZEL 1969 = H. Menzel, Antike Lampen im Römisch-Germanischen Zentralmuseum zu Mainz. Katalog 18, Mainz 1969.

MERCADÉ 1960 = J. Mercadé, Musée de Délos, Paris 1960.

MEULENAERE - MACKAY 1976 = H. De Meulenaere - P. Mackay, Mendes II., Warminster 1976.

MICHALOWSKI etc. 1938 = K. Michalovski - J. de Linage - J. Manteffele - J. Saine Fare Garnot, Fouilles Franco-polonaises. Rapports II. Tell Edfou 1938, Le Caire 1938.

MICHALOWSKI etc. 1950 = K. Michalovski -Ch. Desroches - J. de Linage - J. Manteffele - M. Zejmo-Zejmis, Fouilles

Franco-polonaises. Rapports III. Tell Edfou 1939, Le Caire 1950

MILNE 1924 = J.G. Milne, A History of Egypt under Roman Rule, London 1924.

MLAKAR 1957 = Š. Mlakar, Neki novi antikni nalazi u Istri (Archäologische Funde auf dem Gebiet Istriens), Jadranski zbornik 2, 1957, 433-464.

MÓCSY 1959 = A. Mócsy, Die Bevölkerung von Pannonien bis zu den Markomannenkriegen, Budapest 1959.

MÓCSY 1990 = A. Mócsy, Vallások, in: A. Mócsy - J. Fitz (éds.), Pannonia Régészeti Kézikönyve, Budapest 1990.

MODONESI 1995 = D. Modonesi, Museo Maffeiano. Iscrizioni e rilievi sacri latini, Roma 1995.

MOGENSEN 1930 = M. Mogensen, La Glyptothèque NY Carlsberg. La Collection Égyptienne, Copenhague 1930.

MORA 1990 = F. Mora, Prosopografia Isiaca, II. Prosopografia storica e statistica del culto isiaco, EPRO 113, Leiden 1990.

MORENZ 1960 = S. Morenz, Ägyptische Religion, Stuttgart 1960.

MORENZ 1961 = S. Morenz, Ägyptische Nationalreligion und sogenannte Isismission. ZDMG 111, 1961, 432-436.

MORENZ 1964 = S. Morenz, Gott und Mensch im alten Ägypten, Leipzig 1964

MORENZ 1975 = S. Morenz, Ägyptische Nationalreligion und sogenannte Isismission (in: Religion und Geschichte des Alten Ägypten. Gesammte Aufstze), Weimar, 1975.

MRÁV 2000 = Zs. Mráv, Der Besuch Caracallas und der Deus Invictus Serapis-Kult in Pannonien, CommArchHung, Budapest 2000, 67-97.

MÜNSTER 1968 = M. Münster, Untersuchungen zur Göttin Isis von Alten Reich bis zum Ende des Neues Reich, Berlin 1968;

MYSLIWIEC 1991 = K. Myśliwiec, Athribis – eine hellenistische Stadt im Nildelta, AW 25, 1991, 35-46.

MYSLIWIEC 1992 = K. Myśliwiec, in: Proceedings of the XXXII International Congress for Asian and North African Studies, Hamburg, August 1986, (1992).

MYSLIWIEC-RAGEB 1992 = K. Myśliwiec– M. A. H. Rageb, Fouilles polono-égyptiennes à Tell Atrib en 1986-1990, ÉT 16, 1992, 393-416.

NACHTERGAEL 1994 = G. Nachtergael, Statuettes en terre cuite de l'Égypte gréco-romaine. Recueil des signatures de coroplasthes in: M.O. Jentel – G. Deschenes-Wagner (éds.), Tranquillitas. Mélanges en l'honneur de Tran tam Tinh = Hier pour aujourd'hui, 7. Québec, 1994

NACHTERGAEL 1995 = G. Nachtergael, Une dédicace à Harpocrate, CdÉ 70/139-140, 1995, 246-251.

NACHTERGAEL 1995a = G. Nachtergael, Terres cuites de l'Égypte gréco-romaine. A propos de quatre Catalogue récents, CdÉ 70, 1995, 254-294.

NAGY 1935 = L. Nagy, Aquincumi múmiatemetkezések, Budapest 1935.

NAGY 1942 = T. Nagy, Vallási élet Aquicumban, Budapest története I/2, Budapest 1942, 386-463.

NICOLE 1908 = G. Nicole, Sphinx, in: DAREMBERG-SAGLIO, 1431-1439.

NIKOLANCI 1963-65 = M. Nikolanci, Jedna izgubljena Izida iz Salone (Une Isis de Salone, perdue), VAHD 65-67, 1963-65, 119-126.

NOLL 1955 = R. Noll, Römerzeitliches Sphinxrelief mit griechischer Weihinschrift aus Ägypten, JÖAI 42, 1955, 67-74.

PARLASCA 1970 = Kl. Parlasca, Zur Stellung der Ternuthis-Stelen – Eine Gruppe römischer Grebreliefs aus Ägypten in Berlin, MDAIK 26, 1970, 173-198.

PEARSON – GOEHRING 1986 = B. A. Pearson - J. E. Goehring (éds.), The Roots of Egyptian Christianity, 1986.

PERDRIZET 1921 = P. Perdrizet, Les terres cuites grecques d'Égypte de la Collection Fouquet, Nancy-Paris-Strasbourg 1921.

PETRIE 1935 = W.M.F.Petrie, Shabtis, London 1935.

PETRIE 1886 = W. M. Fl. Petrie, Naucratis I., EEF 3, London 1886.

PETRIE 1910 = W. M. Fl. Petrie, Memphis III., ERAP 1910, London 1910.

PETRIE 1905 = W. M. Fl. Petrie, Roman Ehnasya (Heracleopolis Magna), Plates and Textes Supplementary to Ehnasya, London 1905.

PETTAZZONI 1954 = R. Pettazzoni, The Monstrous Figure of Time in Mithraism, in: Essays on the History of Religion, Leiden 1954.

PHILIPP 1972 = H. Philipp, Terrakotten aus Ägypten. Bilderheft der Staatlichen Museen Preussischer Kulturbesitz Berlin, Heft 18/19, Berlin 1972.

PIANKOFF 1968 = A. Piankoff, The Pyramid of Unas, texts translated with commentary, Princeton NJ 1968.

PHAROS 1995/96 = Pharos. Exibition Catalogue, Zagreb 1995/96.

PLANK 1991 = D. Plank, Das römische Walheim. Ausgrabungen 1980-1988. Archäologische Informationen aus Baden-Württemberg 18, Stuttgart 1991.

PÓCZY 1964 = K. Póczy, Újabb aquincumi múmiasír, Arch. Ért. 91, 1964, 167-191.

PÓCZY 1966 = K. Póczy, Festett férfiportré egy aquincumi múmiasírból (Gemaltes Männerportrait aus einem Mumiengrab in Aquincum), Arch. Ért. 93, 1966, 272-277.

PÓCZY 1968 = K. Póczy, Gemaltes Mumienporträt aus einem Mumiengrab in Aquincum. in: Studien zur Geschichte und Philosophie des Altertums, Budapest 331-337.

PÓCZY 1991 = K. Póczy, Ein Isis-Relief aus Aquincum, in: S. Palágyi (éd.), Acta der X. Kongr. des röm. Kunstschaffens, Veszprém 1991.

PÓCZY 1997-1998 = K. Póczy, What the mummy burials in late Roman Pannonia reveal, Antaeus. Commumicationes 24, 1997-1998, 420-442.

POPA 1983 = Al. Popa, Cîteva consideraţii referitoare la divinităţile egiptene şi microasiatice în epigrafia Daciei romane, Apulum XXI 1983, 71-79.

POSENER 1960 = G. Posener, De la divinité du pharaon, Paris 1960.

POULIN 1987 = S. Poulin, Harpocrate. Essai d'interprétation des représentations d'Harpocrate sur les statuettes et les

lampes en terre cuite, thése Quebec 1987.
PRIJATELJ 1954 = K. Prijatelj, Einige hellenistische Elemente in der Skulptur des antiken Salona, Archeologia Iugoslavica 1, 1954, 29-35.
PUSCHI 1914 = A. Puschi, Nesazio. Scavi degli anni 1906, 1907 e 1908, AMSIA 30, 1914, 1-75.
REICHEL 1893 = W. Reichel, Beschreibung den Sculpturen in Augustustempel in Pola, AEM 16, 1893, 1-13.
REINACH 1913 = A. Reinach, Catalogue des Antiquités égyptiennes, recuellies dans les fouilles de Koptos en 1910 et 1911, exposées au Musée Guimet de Lyon, Chalon-sur-Saone 1913.
REINACH 1910 = S. Reinach, Répertoire de la Statuaire grecque et romaine, II., Paris 1909; IV., Paris 1910.
RENARD 1968 = M. Renard, Sphinx à masque funéraire, Apulum VII/1, 1968, 273-305.
RENDIĆ-MIOČEVIĆ 1989 = D. Rendić-Miočević, Nekoliko monumentalnih stela s portretima iz stare Liburnije (Quelques stèles funéraires monumentales ornées de portraits, de la Dalmatie du Nord), Iliri i anticki svijet (Les Illyriens et le monde antique), Split 1989, 571-588.
RIZZO 1932 = G.E. Rizzo, Prassitèle, Milano-Roma 1932.
ROEDER 1937 = V. Roeder, Ägyptische Bronzewerke, Glückstadt-Hamburg-New York 1937.
ROEDER in RE = G. Roeder, Sarapis, in: RE I A2, 1920, 2394-2426.
RÓMER 1869 = Rómer F, Archeológiai Értesítő 1, 1869, 60.
RONCHI 1974 = G. Ronchi, Lexikon Theonymon rerumque sacrarum et divinarum ad Aegyptum pertinentium quae in papyris ostracis titulis graecis latinisque in Aegypto repertis laudantur. Testi e documenti per lo studio dell Antichita, XLV,1, Milan 1974.
ROULLET 1972 = A. Roullet, The Egyptian and Egyptianizing Monuments of Imperial Rome, EPRO 20, Leiden 1972.
ROUSSEL 1916 = P. Roussel, Les cultes égyptiens à Délos du IIIe au Ier siècle av. J.-C., Nancy 1916.
RÜHFEL 1984 = H. Rühfel, Das Kind in der Griechischen Kunst. Von der minoisch-mykenischen Zeit bis zum

Hellenismus, Mainz 1984.
SACKEN 1871 = E. Sacken, Die antiken Bronzen des k.k. Münz- und Antiken Kabinets in Wien, I., Wien 1871.
SALDITT-TRAPPMANN 1970 = R. Salditt-Trappmann, Tempel der ägyptischen Götter in Griechenland und an der Westküste Kleinasiens, EPRO 68, Leiden 1970.
SEGALL 1938 = B. Segall, Katalog der Goldschmiede-Arbeiten. Museum Benaki Athen, Athen 1938.
SCHOTT 1968 = S. Schott, Falke, Geier und Ibis als Krönungsboten, ZÄS 95, 1968, 54-65.
SCHREIBER 1894 = Th. Schreiber, Die alexandrinische Toreutik I. Abhandlungen der Philologisch-Historischen Klasse der Königlich Sächsischen Geselschaft der Wissenschaft, Leipzig 1894, 271-480.
SCHMIDT 1911 = De Graesk-Aegyptiske Terrakotten i Ny Carlsberg Glyptotek, Copenhague 1911.
SCHMIDT 1994 = E. Schmidt, Martin-von-Wagner-Museum der Universität Würzburg, Katalog der antiken Terrakotten, Teil I. Die figürlichen Terrakotten, Mainz 1994.
SCHWARZ-WILD 1950 = J. Schwartz - H. Wild, Qasr-Qarun/Dionysias I. 1948, Le Caire 1950.
SELEM 1959 = P. Selem, Egipatska božanstva u Arheološkom muzeju u Splitu (Divinités égyptiennes au Musée archéologique de Split), VAHD 61, 1959 (1963.), 94-110.
SELEM 1963-1965 = P. Selem, Egipatski šauabtiji iz zbirke Carrara-Bratanić (Les shaouabtis égyptiens de la collection Carrara-Bratanić), VAHD 64-67, 1963-65, 113-117.
SELEM 1979 = P. Selem, Nekoliko zapažanja o difuziji egipatskih kultova po rimskoj provinciji Dalmaciji (Quelques remarques concernant la diffusion des divinités égyptiennes en Dalmatie), VAHD 72-73, 1979, 79-92.
SELEM 1970 = P. Selem, Stanje istraživanja sfinga carske palače u Splitu (L' état de recherches concernant les sphinx de palais de Dioclétien à Split), Acta praehistorica et antiqua. Miscellanea G. Novak dicata, Zagreb 1970, 633-656.
SELEM 1971 = P. Selem, Boginja s tisuću imena, HZB 1, 1971, 291-332.

SELEM 1972 = P. Selem, Egipatski bogovi u rimskom Iliriku (Document relatifs aux cults égyptiens dans l'Illyricum romain), GCBI IX/7, 1972, 5-104.

SELEM 1994 = P. Selem, La découverte d'une statuette de Nefertoum à Hvar et le problème de la diffusion des cults égyptiens dans les îles de la Dalmatie moyenne, in: LECLANT 1994, 455-460.

SELEM 1997 = P. Selem, Izidin trag. Egipatski kultni spomenici u rimskom Iliriku (Sur les traces d'Isis. Les monuments cultuels égyptiens sur le territoire de l'Illyricum romain), Split 1997.

SENA CHIESA 1997 = G. Sena Chiesa, Iside in età romana: le testimonianze dei materiali, in: ARSLAN 1997, 151-159.

SIMÓN 1990 = F. M. Simón, Iconografia y propaganda ideológica. Júpiter Amón y Medusa en los foros imperiales, Collection Latomus 209, Leiden 1990.

SINN 1977 = U. Sinn, Staatliche Kunstsammlungen Kassel. No.8, Antike Terrakotten, vollständiger Katalog, Kassel-Wilhelmshöhe 1977.

SIST 1997 = L. Sist, L'Iseo-Serapeo Campense, in : ARSLAN 1997, 297-305.

SKUPINSKA-LOVSET 1978 = Ilona Skupinska-Lovset, The Ustinov Collection. Terracottas. Universitetsforlaget, Oslo-Bergen-Tromsó 1978.

SPEELERS 1923 = L. Speelers, Les figurines funéraires Egyptiennes, Bruxelles 1923.

STARK 1997 = Rodney Stark, Der Aufstieg des Christentums, Weinheim 1997.

STEUBEN 1973 = H. Steuben, Der Kanon des Polyklet. Doryphoros und Amazone, Tübingen 1973.

SUIC 1965 = M. Suić, Orijentalni kultovi u antičkom Zadru (Orientalische Kulte im antiken Zadar), Diadora 3, 1965, 91-128

SWOBODA 1964 = E. Swoboda, Carnuntum, seine Geschichte und seine Kunst, Graz-Köln 1964.

SZENTLÉLEKY 1959. = T. Szentléleky, Die Ausgrabungen des Isis-Heiligtums von Savaria, Acta Ant. VII/1-3, 1959, 195-200.

SZENTLÉLEKY 1960 = T. Szentléleky, A szombathelyi Isis-szentély (Das Isis-Heiligtum in Savaria), in: Műemlékeink, Budapest, 1960.

SZENTLÉLEKY 1965 = T. Szentléleky, Az Iseumok kialakulása Egyiptomban és hatásuk Savariára. Savaria 3, 1965, 97-114.

SZENTLÉLEKY 1967 = T. Szentléleky, Architektonische Herausbildung und Entwicklung der Iseen in Ägypten, und ihre Auswirkungen in Pannonien, Acta Ant. 15, 457-466.

SZENTLÉLEKY 1978-79 = T. Szentléleky, Das Isis-Heiligtum von Savaria und die Isis-Noreia-Heiligtümer in Noricum, Schild und Steier 15/16, 1978-79, 179-188.

SZENTLÉLEKY 1990 = T. Szentléleky, Iseum – A szombathelyi Izisz szentély, Szombathely 1990.

SZENTLÉLEKY 1991 = T. Szentléleky, A Püspökkert-Járdányi Paulovics István Romkert a III-IV. században (= Der Járdányi Paulovics István Ruinengarten im III-IV. Jahrhundert), Savaria 20/1, 1991.

TAKÁCS 1995 = S. A Takács, Isis and Sarapis in the Roman World, RGRW 124, Leiden 1995.

THOMAS 1968 = E.B. Thomas, Goldring mit einer Isis-Sothis-Darstellung im ungarischen Nationalmuseum, AV 19 (1968) 119-127.

THOMPSON 1973 = Do. Burr Thompson, Ptolemaic Oinochoai and Portraits in Faience. Aspects of the Ruler-Cult, Oxford 1973.

THOMPSON 1965 = Do. Burr Thompson, Three Centuries of Hellenistic Terracottas, Hesperia 34, 1965, 33-71.

TIBILETTI 1997 = T. Tibiletti, La festa del "Navigium Isidis", in: ARSLAN 1997, 658-659.

TIBILETTI 1997a = T. Tibiletti, Il sistro, in: ARSLAN 1997, 66-661.

TONKOVIĆ 1991 = S. Tonković, Brončana statua Izide-Fortune iz Prološca, Imotski, Obavijesti HAD XXIII/3, 1991, 53-55.

TÓTH 1980 = E. Tóth, Silvanus Viator, Alba Regia 18, 1980, 91-103.

TÓTH 1983-1984 = E. Tóth, Ein Goldring mit Rorträtgemme

eines Julisch-Claudischen Prinzen (Germanicus), RÖ, 1983-1984, 315-336.

TÓTH 1985 = E. Tóth, Római gyűrűk és fibulák. Évezredek évszázadok kincsei 3, Budapest 1985.

TÓTH 1976 = I. Tóth, Marcus Aurelius esőcsodája és az egyiptomi kultuszok, AT 23, 1976, 45-51.

TÓTH 1977 = I. Tóth, Ein "Viergötterstein" mit ägyptischen Göttern aus Savaria (ad CIL III 10908 = SIRID 662), Studia Aegyptiaca III, Budapest 1977, 131-148.

TÓTH 1978 = I. Tóth, Négyalakos szoborbázis a savariai Iseumból (Ad CIL III. 10908. = RIU 15.), Vasi szemle 32/4, 1978, 585-596.

TÓTH 1996 = I- Tóth, Ovidius és Savaria – Egy legenda, in: I. Lengvári (éd.), In memoriam Barta Gábor. Tanulmányok Barta Gábor emlékére, Pécs 1996, 1-17

TÓTH 1996-1997 = I. Tóth, A savariai Iseum kutatásának eredményei és feladatai, Savaria 23/3 (1996-1997) 1998, 329-352.

TÓTH 1998 = I. Tóth, Hic numina adsunt..., M. Kiss - I. Lengvári (éds.), A népvándorláskor fiatal kutatóinak VII összejövetele. Pécs 1996. szept. 27-29, Pécs 1998, pp. 9-18,

TÖRÖK 1995 = L. Török, Hellenistic and Roman Terracottas from Egypt – Monumenta Antiquitatis extra Fines Hungariae Reperta IV., Roma 1995.

TRAN TAM TINH 1970 = V. Tran tam Tinh, Isis et Serapis se regardant, RA I, 1970, 55-80.

TRAN TAM TINH 1971 = V. Tran tam Tinh, Le culte des divinités Orientales à Herculaneum, EPRO 17, Leiden 1971.

TRAN TAM TINH 1973 = V. Tran tam Tinh, Isis lactans. Corpus des monuments gréco-romains d' Isis allaitant Harpocrate, EPRO 37, Leiden 1973.

TRAN TAM TINH 1983 = V. Tran tam Tinh, Sérapis debout, EPRO 94, Leiden 1983.

TRAN TAM TINH 1988 = V. Tran tam Tinh – B. Jaeger – S. Poulin, Harpocrates, in: LIMC IV, 1988, 415-445, pl. 242-266.

UHLENBROCK 1990 = J. P. Uhlenbrock, The Coroplast's Art. in: J. P. Uhlenbrock (éd.), Greek Terracottas of the Hellenistic World, New York 1990.

URANIĆ 2001 = I. Uranić, in: Egipatska religija i antička Istra, Pula 2001, 22-28.

VANDERLIP 1972 = V. F. Vanderlip, The Four Greek Hymns of Isidorus and the Cult of Isis, American Studies in Papyrology 12, 1972, 1-108.

VANDIER 1971 = J. Vandier, Nouvelles acquisitions. Musée de Louvre. Département des Antiquités Égyptiennes, La Revue du Louvre et des Musée de France 2, 1971, 95-106.

VARGA-LÁSZLÓ-KOVÁCS, 1987 = E. Varga – P. László – P. Kovács, Egy Izisz papnő koporsója magyar gyűjtményben – Le sarcophage d'une prêtresse d'Isis dans la collection hongroise, BMHBA 68-69, Budapest 1987, 3-52 et 203-224, figs. 1-41.

VARGA 1994 = E. Varga, Le passé, le présent et l'avenir de la Collection égyptienne de Budapest, BSFE 131, 1994, 19-36.

VERZAR 1978 = M. Verzar, Aventicum II. Un temple du culte impèrial – Contexte stylistique et typologique du décor architectonique, Lausanne 1978.

VIDMAN 1965 = L. Vidman, Die Isis und Serapisverehrung im 3. Jahrhundert u. Z., Neue Beiträge zur Geschichte der alten Welt, II, Römisches Reich, Berlin 1965.

VIDMAN 1970 = L. Vidman, Isis und Sarapis bei den Griechen und Römern. Epigraphische Studien zur Verbreitung und zu den Trägern des ägyptisches Kultes, Reliogionsgeschichtliche Versuche und Vorarbeiten, XXIV, Berlin 1970, 125-138.

VIDMAN 1981 = Rec. par L. Vidman: G. Hölbl, Zeugnisse ägyptischer Religionsvorstellungen für Ephesus, BiO 36, 1979, 322-323.

VLASSA 1980 = N. Vlassa, Sfinxul de bronz de la Potaissa, Potaissa. Studii şi comunicări 2, Turda 1980, 133-153.

VOGT 1924 = J. Vogt, Expedition von Ernst von Sieglin. Ausgrabungen in Alexandria, II. Die griechisch-ägyptische Sammlung E. von Sieglin, 2. Terrakotten, Leipzig 1924.

VOLLENWEIDER 1984 = M-L. Vollenweider, Portrais d'enfants en minature de la Dynastie des Ptolémées, in: Studi e materiali di Istituto di Archeologia. Università di Palermo 5. Alessandria e il mondo ellenistico-romano. Studi in onore di Achille Adriani, Roma 1984, 363-377.

VOLLENWEIDER 1995 = M-L. Vollenweider, Camées et intailles. Tome I: Les portraits grecs du Cabinét des médailles, Paris 1995.

VORBECK 1980 = E. Vorbeck, Zivilinschriften aus Carnuntum, Wien 1980.

WALKER-HIGGS 2000 = S. Walker - P. Higgs (éds.), Cleopatra. Regina d'Egitto, Milano 2000.

WALKER-HIGGS 2001 = S. Walker - P. Higgs (éds.), Cleopatra of Egypt. From History to Myth. London 2001.

WEBER 1914 = W. Weber, Die Ägyptisch-Griechischen Terrakotten. Mitteilungen aus der Ägyptischen Sammlung, Bd. 2, Berlin 1914.

WEINGAARDEN 1958 = W.D. van Wijngaarden, De Grieks-Egyptische terracottas in het Rijksmuseum van Oudheden, Leiden 1958.

WEISSHAUPL 1901 = R.Weisshäupl, Zur Topographie des alten Pola, JÖAI IV, 1901, 179-182.

WESSETZKY 1959 = V. Wessetzky, die Probleme des Isis-Kultes in Ober-Pannonien, Acta Arch. 11, 1959, 265-282.

WESSETZKY 1959b = V. Wessetzky, A felsőpannóniai Isiskultusz problémái (Sur les problèmes de la culte isiaque en Pannonie supérieure), Arch. Ért. 86, 1959, 20-31.

WESSETZKY 1961 - Die Ägyptischen Kulte zur Römerzeit in Ungarn, EPRO 1, Leiden 1961.

WESSETZKY 1965 = V. Wessetzky, Ägyptische Götter in Pannonien, Savaria 1965, 141-146.

WESSETZKY 1989 = V. Wessetzky, Isis és Osiris Pannoniában, Budapest 1989.

Wien 1994 = Ägyptomanie. Ägypten in der europäischen Kunst. 1730-1930, Wien 1994.

WILD 1981 = R. A. Wild, Water in the Cultic Worship of Isis and

Serapis, EPRO 87, Leiden 1981.
WILD 1984 = R.A. Wild, The Known Isis-Sarapis Sanctuaries from the Roman Period, in: ANRW II. 17/4, 1984, 1739-1851.
WINTER 1903 = Fr. Winter, Typen der figürlichen Terrakotten, II., Berlin-Stuttgart 1903.
R.E. Witt 1971 = Isis in the Greco-Roman World, London 1971.
ŽANIĆ-PROTIĆ 1988 = J. Žanić-Protić, Antička brončana plastika iz Arheološkog muzeja u Splitu I. (Roman bronze Sculpture from the Archaeological Museum in Split I.), VAHD 81, 1988, 21-32.

1-2.Portrait of Marcus from Dunaszekcső 3. Portrait of Marcus 4-5. Marcus' portrait as a teenager 6. Marcus' portait at the age of 30.

Pl. 2

1. The entire statue monument and the Doryphoros 2. Close shot of the head of the statue monument 3. Two Serapis monuments 4. Rain Miracle on the relief of the Aurelian column in Rome

1. Marble head of Hathor from Pola 2. Tombstone of Obellia Maxima from Pola

Pl. 4

1. Isis's votive altar from Galižana near Pola, drawings 2. Its front side 3. Its right lateral side

1-2. Isis's votive altar from Galižana near Pola. The left lateral side.

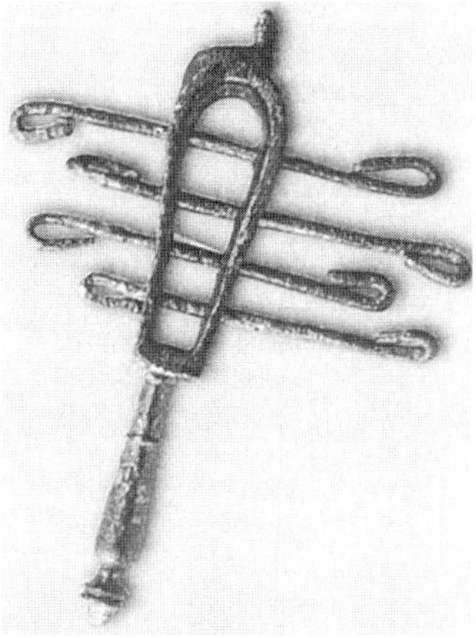

1. Bronze sistrum from Nesactium 2. Bronze statuette of Isis-Fortuna from Salvore 3. Marble statue of Isis from Aenona

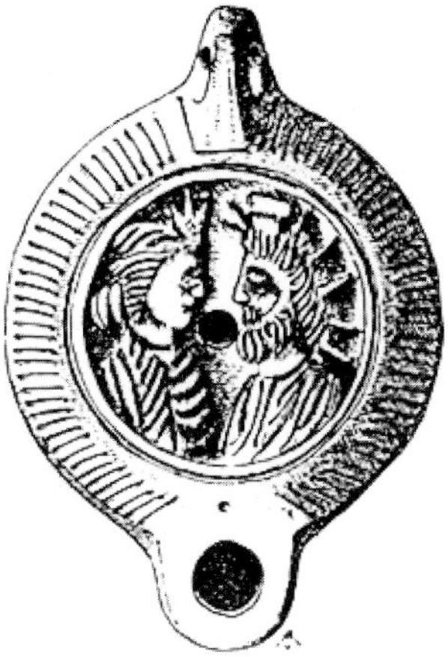

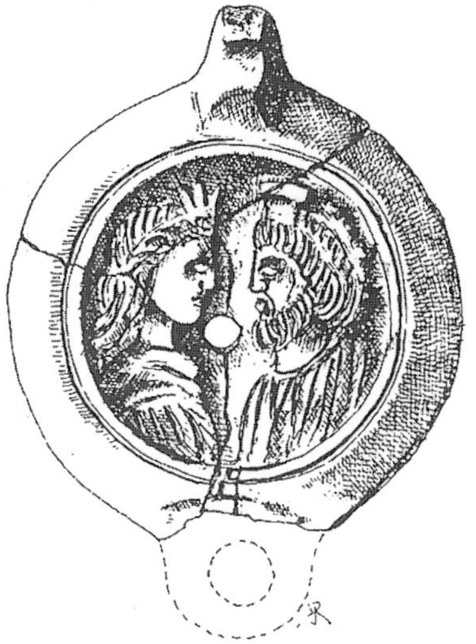

1-2. Lamps with the representation of Isis and Serapis from Aenona 3. Lamp with the representation of Harpocrates from Aenona

Pl. 8

Votive altar dedicated to Isis and Serapis,Liber and Libera from Iader: 1. front side 2. left lateral side 3. right lateral side 4. back side

1. Press-model (negative and positive) with the representation of Serapis from Iader 2. Votive altar dedicated to Isis from Varvaria

1. Bust of Isis from Salona 2. Bust of Serapis from Salona 3. Applique or phalera with the figure of Jupiter-Amon from Salona 4. Gem representing Anubis from Salona 5. Wedjat-eye amulet from unknown site 6. Gem representing Isis and Serapis from Pharos

1-2. Cover of the sarcophagus of Aurelius Satrius and Aurelia Maxima from Salona 3. Its left medaillon representing Aurelius Satrius 4. Its right medaillon representing Aurelia Maxima

Pl. 12

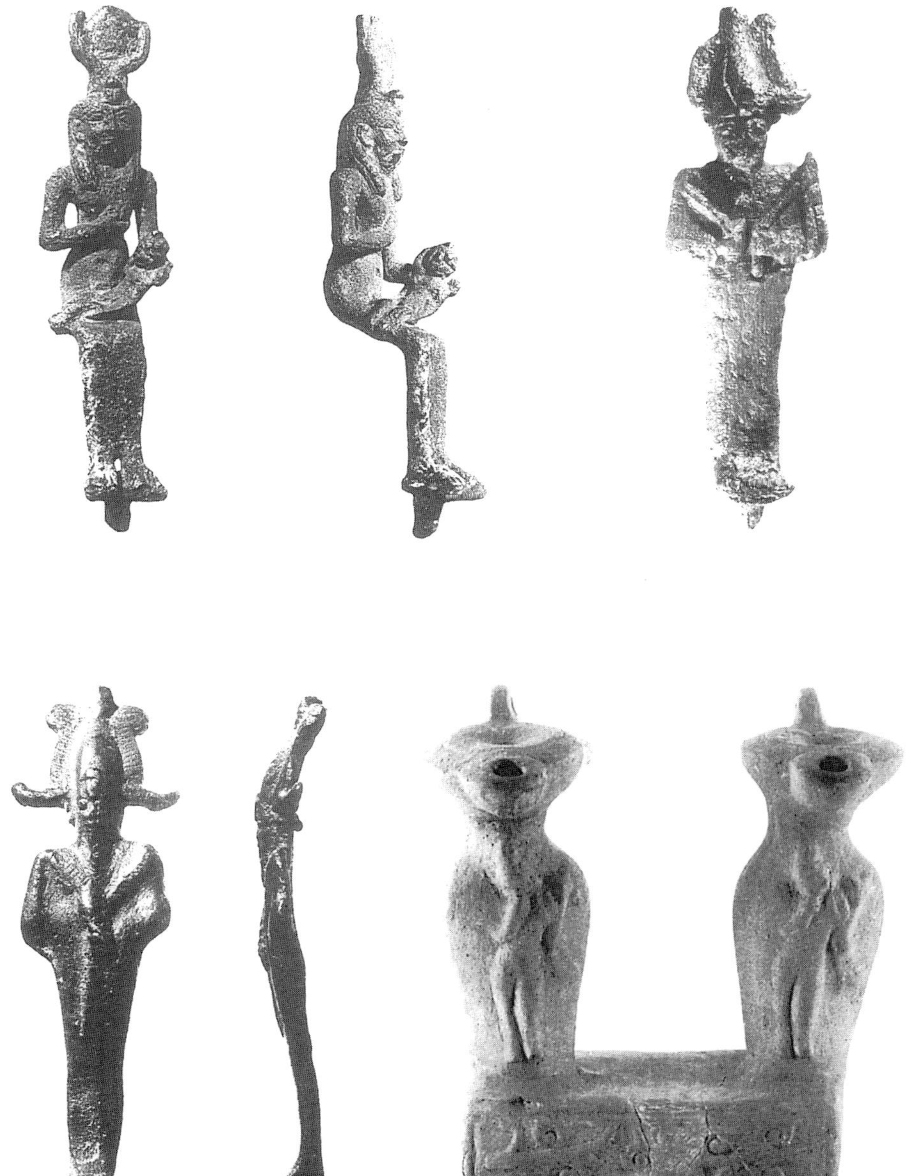

1-2. Bronze statuette representing Isis lactans from Salona 3-4. Bronze statuette representing Osiris from Salona 5. Lamp with the representation of two Harpocrates from Salona

1. Sphinx on the Peristyle of the Diocletian's palace in Split 2. Sphinx of Amenophis III from the Diocletian's palace in Split 3. Sphinx in front of the Temple of Jupiter in the Diocletian's palace in Split.

Pl. 14

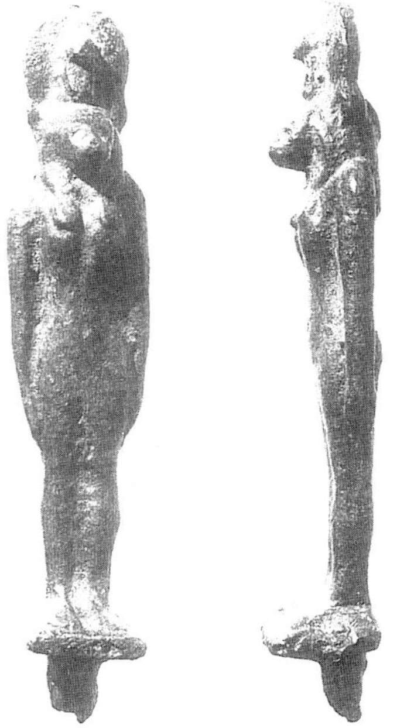

1. Bust of Isis from Narona 2. Egyptian head from the Diocletian's palace in Split 3. Bronze statuette of Nefertum from Pharos 4. Bronze statuette of Osiris from Pharos 5. Bronze statuette of Sekhmet from Epidaurum

1. Bronze statuette of Isis-Fortuna from the territory of Lika 2. Bronze vessel in the shape of Bes from Oklaj 3. Bronze statuette of Isis-Fortuna from Proložac near Imotski 4. Monotith limestone blocks of the upper part of an altar enclosure with the heads of Jupiter-Amon from Pola

1-4. Monotith limestone blocks of the upper part of an altar enclosure with the heads of Jupiter-Amon from Pola

1-3. Two pilasters with the heads of Jupiter-Amon and one pilaster with the head of Medusa from Iader

Pl. 18

Terres cuites représentant Harpocrate-au-faucon-sur le pilier 1.-4. au Caire et 5.-6. aux Musées Royaux d'Art et d'Histoire, Bruxelles

1-4. Terres cuites représentant Harpocrate-au-faucon-sur le pilier à Brooklyn

Pl. 20

Terres cuites représentant Harpocrate-au-faucon-sur le pilier au Musée British

Inscription sur le pilier de la statuette du Musée British

Die Kanne von Egyed – 1. Das Verhältniß des Schnabels zum Götterfries 2. Das Verhältniß der Schweißnähte zum Götterfries

1-2. A coin representing the Iseum on the Campus Martius in Rom
3. Isis liturgy in a fresco in Herculaneum

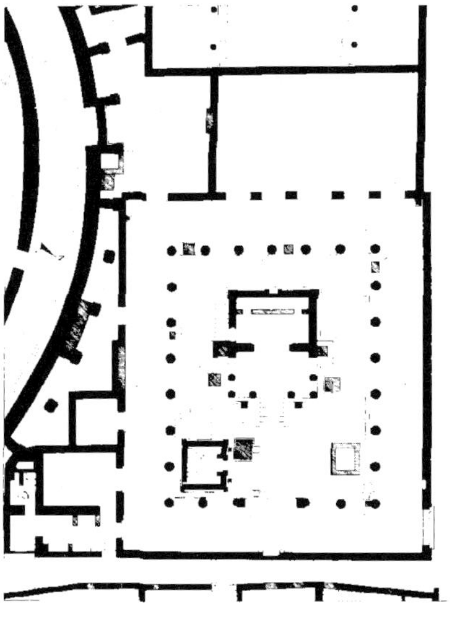

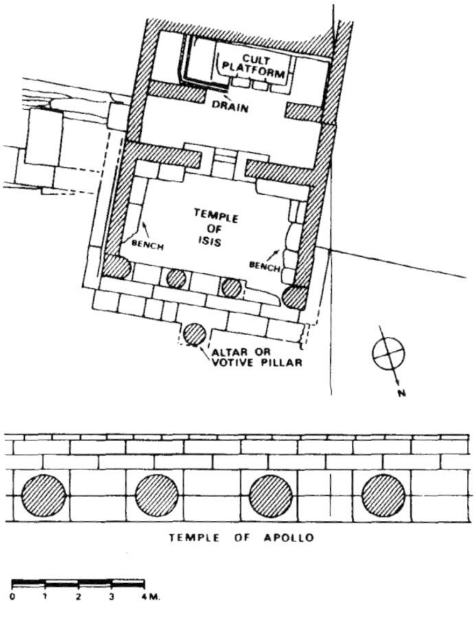

1. The Iseum in Pompei by Piranesi
2. The Iseum in Pompei
3. The Iseum in Cyrene

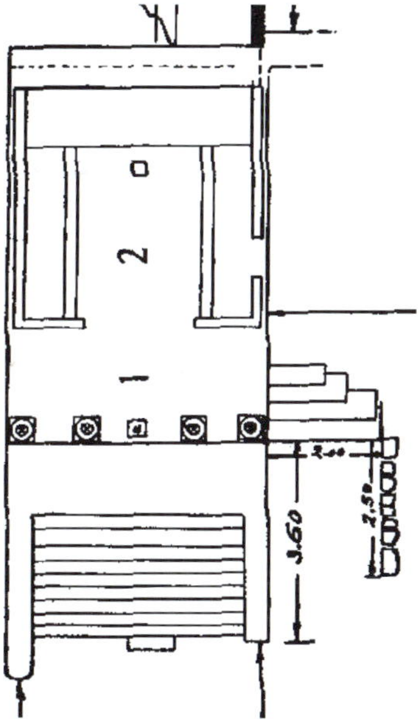

1. The Iseum in Thabessa, archiv.
2-3. The Serapeum in Ras el-Soda. Ground-plan and view.

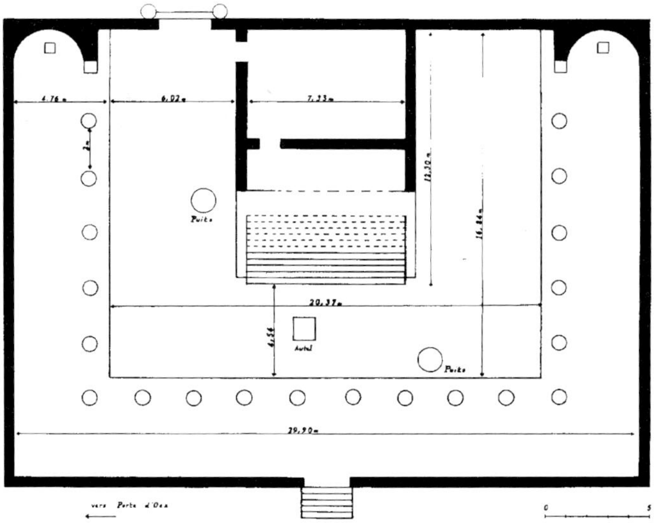

1-2. The Serapeum in Leptis Magna. Ground-plan and view

1-2. The ring decorated with plastic bust of Sarapis in the Hungarian National Museum 3-4. Bronze Sarapis ring from Dieburg

Gilded bronze ring with busts of Isis and Sarapis in the British Museum

1. Isis Relief, Savaria 2. Inschrift des Altars von Aquincum 3. Porträt von einem ägyptischen Gott, Carnuntum

1. Serapis-Isis relief, (inv. no. 64.11.166) Aquincum 2. Isis relief, (inv. no. 64.10.187) Aquincum 3. Isis Pharia auf einer Münze

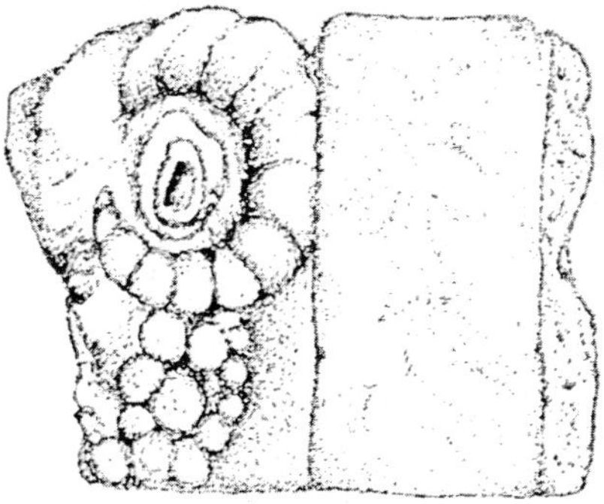

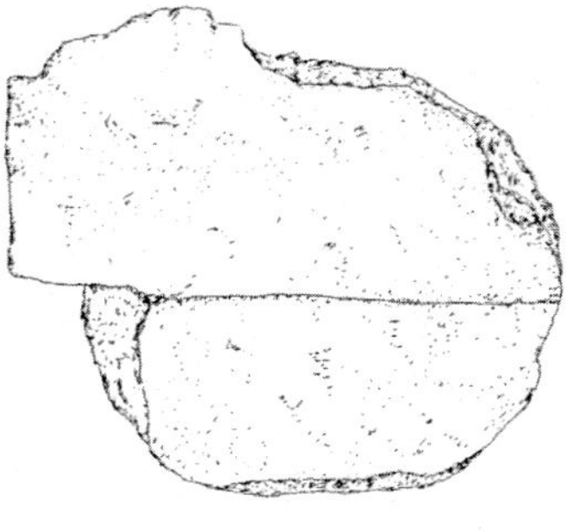
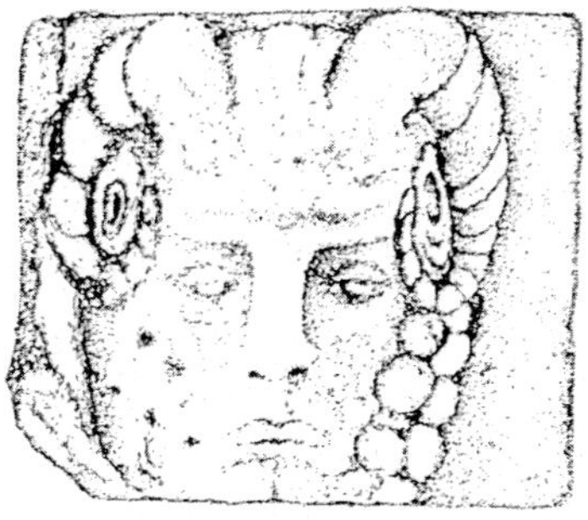

1. Säulenkapitell geziert mit Ammonkopf, Aquincum 2-5. Konsole dekoriert mit dem Kopf des Zeus-Amun, (inv. no. 81.8.15.) Aquincum

1. Prozessionsszene mit einem Priester in der Mitte 2. Der Priester mit dem Anubis-Mask, Savaria —-Iseum